THE RELIABILITY FRAMEWORK

Engineering for reliability

THE RELIABILITY FRAMEWORK • PART 8

Engineering for reliability

FMEA, RCM, Maintainability and LCC for Projects

MARTIN VAN DEN HOUT

VULCANUS PRESS
FABRICANDO FIT FABER

Colophon

Title:	Engineering for Reliability: FMEA, RCM, Maintainability, and LCC for Projects
Series:	The Reliability Framework: Nine Guides to Flawless Asset Performance
Volume:	8
Author:	Martin van den Hout
Photos:	Depositphotos.com
ISBN:	9789083643199
Edition:	First edition: April 2026

All measurements in this book are metric. For most examples, the exact units are not critical to understanding the principle being explained. Where precision matters both metric and imperial values are provided.

Liability and Disclaimer

The information in this book has been compiled with great care, drawing on 35 years of practical experience in maintenance and reliability engineering. The goal is to provide maintenance professionals with practical tools for improving their maintenance processes.

Neither the author Martin van den Hout, nor Vadeo BV, nor Vulcanus Press can accept liability for the consequences of actions based on this book. Every business situation is unique and requires its own assessment of risks, costs, and benefits.

When implementing the described methods, you should take into account the specific circumstances of your organization, applicable laws and regulations, safety requirements, and manufacturer guidelines. When in doubt, consult specialists and make responsible decisions based on thorough risk analysis.

This book is intended as a practical guide and does not replace customized professional advice, adequate employee training, or sound engineering judgment.

Foreword

Anyone who works daily to keep their assets reliable knows how exceptional that profession is. Behind every running pump, every production line that keeps operating, and every network that keeps functioning, stands a skilled professional with knowledge, dedication, and an unstoppable curiosity for improvement. The maintenance professional is the silent force that makes technology perform, monitors processes, and moves organizations forward — often without anyone noticing, until something goes wrong.

Yet the world of maintenance is changing faster than ever. From TPM to Lean, from Asset Management to Reliability Engineering - the terminology shifts, but the core remains the same: reliability starts with people who know what they're doing, and who dare to learn, share, and improve.

This book is an invitation to you - the professional who looks beyond repairing what is broken. It is a compass for those who view maintenance not as a cost, but as the engine of continuity, safety, and value creation. Whether you work in industry, the public sector, or at a high-tech company, the principles shared here are universal: structured thinking, craftsmanship, and working together toward reliability.

Let this book inspire you to look beyond today. See every maintenance moment as an opportunity to learn, every problem as a reason to improve, and every asset as a system that can do better — because you want to be better.

My admiration goes out to Martin van den Hout, author and knowledge custodian with decades of experience in the maintenance and Asset Management field.

His sharp insights, thorough knowledge, and passion for sharing expertise have made an invaluable contribution to this work. Thanks to his ability to make complex subject matter clear and accessible, his contributions form a valuable bridge between theory and practice.

Because, ultimately, maintenance is not just a profession: It is a mindset!

Ellen den Broeder-Ooijevaar
Manager NVDO
(Dutch National Maintenance Society)

Table of Contents

Foreword **vi**
Introduction **xi**
1 The Cost of Unreliability **19**

1.1 The Iceberg Principle: What You Don't See 19
1.2 Reliability as a Competitive Advantage 21
1.3 The 80/20 Rule of Reliability Engineering 23
1.4 Summary: The Core Lessons 25

2 From Initiative to Specification **27**

2.1 The Dangers of Drawing Conclusions Too Quickly 28
2.2 Systematic Problem Analysis: The 5 Whys Technique 29
2.3 Stakeholder Analysis: Who Has a Stake in the Solution? 29
2.4 Early Equipment Management 31
2.5 Functional versus Technical Specifications 34
2.6 Quantitative Reliability Objectives 35
2.7 Class vs. Quality: Specifying Installation Quality 36
2.8 Identifying Boundary Conditions and Constraints 38
2.9 Future-Proofing and Flexibility 39
2.10 The Specification Document: Structure and Content 40
2.11 Validation of Specifications 41
2.12 Common Mistakes in the Specification Phase 42
2.13 RAMS Methodology 46
2.14 Project Phases with Approval Criteria 48
2.15 Resource Planning for Reliability Engineering 48
2.16 Summary 49

3 Criticality Analysis **51**

3.1 When and Why 51
3.2 Sarah's Story: Three Days That Changed Everything 52
3.3 The Risk Matrix 53
3.4 Types of Risks 53
3.5 The Criticality Ranking: How You Do It 56
3.6 From Score to Strategy: The Consequences 59
3.7 The Business Case Impact: Concrete Numbers 63
3.8 Efficiency and Pragmatism 64
3.9 Quality Assurance of the Process 65
3.10 Summary: The Core Lessons 66

4 FMEA – Systematic Failure Analysis 69

4.1 FMEA: The Core of Reliable Design 69
4.2 The Seven Core Questions of FMEA 70
4.3 FMEA Team Composition and Process 73
4.4 From FMEA to Design Improvements 75
4.5 Practical FMEA Implementation 76
4.6 Common Mistakes in FMEA 78
4.7 Summary: The Core Lessons 79

5 Design for Maintainability 81

5.1 The Story of Site Manager Pedro Vargas 81
5.2 Maintainability as a Design Criterion 82
5.3 The Nine Elements of Downtime 83
5.4 The Five Design Principles for Maintainability 84
5.5 Documentation and Knowledge Transfer 88
5.6 Safety and Regulations 90
5.7 Common Mistakes and Pitfalls 90
5.8 Summary: The Core Lessons 91

6 RCM in Projects 95

6.1 Why RCM in Projects? 95
6.2 RCM Methodology in Brief 96
6.3 Selective Application of RCM 100
6.4 Practical Application in Projects 101
6.5 Results, Lessons, and Recommendations 102
6.6 Summary 102

7 Life Cycle Costing 105

7.1 The Problem of Short-Sighted Decision-Making 105
7.2 LCC Methodology: Cradle-to-Grave Thinking 107
7.3 Practical Tools for LCC Analysis 109
7.4 Input for Calculations and Data Challenges 110
7.5 Sandra's complete LCC equation 111
7.6 Remaining Service Life Analysis and Replacement Strategies 113
7.7 Digital Support for LCC 120
7.8 Summary: The Power of Systematic Thinking 120

8 Building the Business Case 123

8.1 ROI Calculation of Reliability Investments 124
8.2 Accounting for Risks in Calculations 126
8.3 A Standard Model for Presenting the Business Case 127
8.4 Summary: The Power of Compelling Business Cases 128

9 Commissioning and Operational Readiness 131

9.1 Commissioning: The Three Aspects Simultaneously 132
9.2 Systematic Planning 132
9.3 Assessment of Operational Readiness 133
9.4 Training and Skills Development 135
9.5 Procedure Development and Documentation 135
9.6 Integrated Testing and Reliability Validation 136
9.7 Reliability Growth and Lessons Learned 137
9.8 Handover and Performance Monitoring 138
9.9 Summary: What Really Works 139

10 Supplier Selection and Procurement 141

10.1 Procurement versus Project Management 141
10.2 Requirements for Dependability Contracts 143
10.3 Training and Development of Competencies 146
10.4 Stability of finances and business continuity 147
10.5 Summary: The Core Lessons 147

11 Asset Portfolio Management 151

11.1 Strategic Pressure from Management 151
11.2 Systematic Portfolio Approach 152
11.3 Identifying Synergies Within the Portfolio 154
11.4 Scenario Analysis for Robust Decisions 155
11.5 The Decision 155
11.6 Summary: From Methodologies to Portfolio Strategy 156

Epilogue: Engineering for Reliability in Perspective 158
Appendix A Extended Scorecard for Supplier Evaluation 162
Figures 168
Glossary — Terminology and Acronyms 169

Introduction

This book is part of a series that answers the question: "What should an organization do to make its technical installations perform as well as possible?"

I asked myself this question 35 years ago, and it has kept me constantly engaged ever since. At the time, I was working in the Maintenance & Engineering department of a Japanese manufacturing company in the Netherlands. In those days, the term "Total Productive Maintenance" (TPM) was on the rise, and more and more people were beginning to realize that simply setting up a good maintenance schedule is not enough to keep a production plant running smoothly.

Within TPM, the emphasis was strongly on involving production operators and improvements within a production department. I realized fairly quickly that this was a major step forward, but still not enough.

There were also many questions within our maintenance department that TPM didn't answer. Questions like: “Which spare parts do we need, and how many?” or “What is the best way to organize the large flow of work within a maintenance department, ranging from jobs that take a minute to projects requiring thousands of man-hours?”

In the past thirty-five years, our field has matured. TPM was incorporated into Lean Manufacturing, and from (semi-) government bodies, more and more attention was paid to technical management, under the name asset management. This ultimately led to the publication of the ISO 55000 series of standards. For the reliable operation of a production plant or other facility, the term ‘Reliability’ emerged as a replacement for the term ‘Maintenance’. After all, reliability is the goal, while maintenance is merely one of the means.

I learned that all these schools of thought strongly resembled one another and that the differences were mainly differences in emphasis, driven by corporate culture and factors such as how many operators are directly involved in production. In the chemical industry their role is different from that in the automotive industry, and in government organizations this role is different again.

Ultimately, many organizations have combined elements from these approaches in whatever way best suited their specific situation. Yet I learned that what applies to a production plant also applies to a network, your car, or a power plant. It doesn't matter.

It always bothered me that there were only two kinds of books about this field. The first kind describes the entire field, but only describes 'what' you should organize. They are therefore very broad, but never go into any depth. The second kind focuses on one aspect, such as workflow control or setting up preventive maintenance, and does go into depth on how to apply that one aspect.
These books have a number of things in common. They are not interconnected with one another, there are very few good ones, they are usually written in foreign languages, and they are always very thick. Often they are also written from the perspective of 'the ideal world,' in which you have an infinite number of engineers or technicians at your disposal and the most fantastic computer programs.

How This Book Is Structured

This book deliberately follows the natural steps you go through when designing and building reliable assets. You can think of it as a journey in six phases, where each chapter builds on the previous one.

The problem understanding (Chapters 1–3) begins with the why.
shows you why reliability is so valuable and how the 80/20 rule means that the most important decisions are made during design. You will learn the true costs of unreliability — often 3–5 times higher than the visible repair costs.

Chapter 2 covers the crucial step from initiative to specification.
Here you will learn how systematic problem analysis using the 5 Whys technique helps you get to the root causes. Stakeholder analysis and Early Equipment Management ensure that all perspectives are taken into account from the beginning. The quantitative reliability targets you formulate here determine the success of your entire project.

Chapter 3 covers criticality analysis as the first analytical step. Here you learn to systematically prioritize by assessing assets on five risk dimensions: safety, environment, delivery reliability, quality, and sustainability. The risk matrix helps you objectively determine which assets deserve extra attention during engineering.

Designing for Reliability (Chapters 4–6) is where theory becomes practice. Chapter 4 lets you systematically analyze failure modes with FMEA. The seven core questions force you to look beyond desired functionality toward all the ways systems can fail. The Risk Priority Number helps you prioritize where improvements have the most impact.

Chapter 5 covers maintainability as a design criterion. The five design principles - accessibility, visibility, standardization, modularity and built-in diagnostics - determining whether your asset truly maintainable looks like in practice. The nine elements of downtime show that active repair time is often only 25% of the total downtime.
In chapter 6, you prepare maintenance strategies with RCM in projects. By systematically analyzing what maintenance is truly necessary, you prevent both over- and under-maintenance. The decision diagrams guide you to the optimal maintenance strategy for each failure mode.

Economically justifying (Chapters 7–8) ensures your projects are financially compelling.
Chapter 7 covers Life Cycle Costing as a methodology to further to consider than purchase price. Sandra's pump comparison shows how systematic LCC analysis prevents suboptimal decisions. The Net Present Value calculations and sensitivity analysis make your analysis robust.

Chapter 8 describes how to build solid business cases that both technically sound as well as financially convincing. The ROI calculation methodology and risk analysis give management the confidence to invest in reliability. You will learn to translate FMEA insights and LCC figures into clear financial presentations.

Realization (Chapters 9–10) transforms plans into reality.
Chapter 9 makes commissioning practical as the phase in which engineering, people and organization come together. Frank's biomass plant illustrates why operational readiness is just as important as technical performance. The systematic planning of five commissioning phases prevents technically perfect installations from failing due to a lack of organizational preparation.

Chapter 10 covers supplier selection beyond traditional purchasing criteria. Anna's scorecard systematizes the evaluation of technical capabilities, operational certainty, and long-term
support. The contractual reliability requirements ensure that warranties deliver real value.

Portfolio Strategy (Chapter 11) looks at the bigger picture. This
chapter shows how asset portfolio management identifies synergies that are not visible in asset-by-asset analysis. Karen's energy plant project shows how an integrated approach creates $1.4 million in additional value by systematically examining portfolio interactions
look.

The epilogue concludes with a reflection on how reliability engineering
is interconnected with the other eight aspects of maintenance management that you need for sustainable top performance.

This series consists of nine books, which together provide the answer to the previously mentioned question: "What should an organization do to make its technical installations perform as well as possible?" The books cover one topic each in a systematic, step-by-step approach. Each book connects to the other books in the series. This way, you can implement a coherent system within your company to allow your assets to perform optimally.

This is the eighth book in this series. The series consists of the following books:

Part 1: Maintenance Management

A comprehensive overview of the entire methodology. A general description of the topics covered in the other eight parts, plus several additional subjects.
ISBN 9789083643120

Part 2: Planning and Scheduling in Maintenance

Organization of the workflow of the maintenance department; including work requests, gate-keeping, work preparation, scheduling, supervision, safety, administration, improvement, and archiving.
ISBN 9789083643137

Part 3: Troubleshooting Breakdowns

The resolution of an individual failure, by a maintenance technician or an operator. Notification, finding the cause, finding a solution, repair, failure reporting, continuous improvement.
ISBN 9789083643144

Part 4: Root Cause Analysis in Maintenance

Reducing the number of failures that occur annually. Registration, statistical analysis, prioritization, root cause analysis, and implementation of solutions.
ISBN 9789083643151

Part 5: Preventive Maintenance

Setting up a preventive maintenance schedule for an entire organization, taking into account the risks and criticality of the individual assets. Translating management requirements for the entire organization into technical requirements for the assets. Mapping possible failure modes, causes, and possible maintenance actions. Economic approaches to determine which maintenance is worthwhile. Methodologies such as RCM, Industrial RCM, and pragmatic approaches.
ISBN 9789083643168

Part 6: Spare Parts Management

What types of spare parts are there? Is it worthwhile to keep them in stock? How many of each type? Organization of the warehouse and key considerations for common types of components.
ISBN 9789083643175

Part 7: Precision Maintenance

Many assets fail prematurely because they have not been assembled with sufficient precision. A technical handbook covering practical topics such as bolt tightening, fits, aligning rotating equipment, assembly with attention to balance, flange assembly, and so forth.
ISBN 9789083643182

Part 8: Engineering for Reliability

80 percent of maintenance costs and technical reliability is determined during design and construction. Project management, Early Equipment Management, involvement of operators and technicians, reliability analyses during design, FAT and SAT, Commissioning, Maintainability.
ISBN 9789083643199

Part 9: Autonomous Maintenance

Which maintenance tasks can you have operators perform, why, and how do you organize that?
ISBN 9789083706306

For a current overview of all books, additional information, and ordering options, visit:

vadeo.nl/books.html

1 The Cost of Unreliability

It was a Tuesday morning in March when production manager Mark at Bacchus Food Group stood before the halted production line of their food processing plant. The main motor of the packaging machine had given out — for the third time that year.

While the technicians got to work, Mark began tallying up the damage. Four hours of downtime meant an initial loss of 4,000 packages, he quickly calculated. But as the day progressed, Mark realized his calculation had been far too optimistic.

The actual costs ran much higher: $1,600 in lost margin (4,000 packages × $0.40), $800 in ongoing fixed costs, $500 in startup losses, $300 in quality problems due to the restart, plus $1,000 in repair costs. The total came to $4,200 for this one failure.

"How can this be?" Mark asked. "I had calculated everything properly, hadn't I?"

This story repeats itself daily in plants, hospitals, power stations, and other organizations all over the world. Managers focus on the purchase price of assets, but systematically underestimate the costs of unreliability. That is precisely why reliability engineering is no longer a technical luxury, but a strategic necessity for every organization that wants to survive in the modern economy.

1.1 The Iceberg Principle: What You Don't See

When you analyze the financial impact of asset failures, you quickly discover that the iceberg is much larger than its visible tip. The direct repair costs — parts and labor — often represent only a small fraction of the actual damage. The real costs are hidden in five areas that many organizations fail to map adequately.

Production Loss: The Biggest Cost Item

During downtime, you not only lose revenue, but fixed costs keep running. Personnel stand idle, yet still have to be paid. Energy is wasted during the restart of processes. In many industries, 1 hour of downtime on the main production line costs more than the annual salary of an experienced technician.

Mark's packaging machine normally produced 1,000 packages per hour with a margin of $0.40 per package. Four hours of downtime therefore meant $1,600 in lost margin. But on top of that came the ongoing costs: labor ($500), energy for other systems ($200), and overhead costs ($100). Total production loss: $2,400.

Quality Problems: The Hidden Multiplier

After an unexpected shutdown, processes must be restarted and recalibrated. Temperatures, pressures, and mixing ratios need to stabilize. The first products after restart often fail to meet quality standards and must be rejected or reworked.

In Mark's case, it took two hours before the machine was running to specification again. During that time, he produced 2,000 packages that did not meet quality requirements — a loss of $300 in material and labor.

Safety Issues: Risks Under Pressure

Under pressure to get back up and running quickly, operators and technicians sometimes take risks. Safety protocols are skipped. Emergency solutions are implemented without thorough risk analysis. Statistics show that the number of workplace accidents increases significantly in the weeks following major failures. 75 percent of all accidents in production companies happen during maintenance activities. Solving failures in particular is dangerous, because it always happens unprepared and operators often don't know exactly what is wrong with the system.

Although Mark's failure fortunately caused no accidents, he was concerned about the way his team had carried out the repair. In their haste to get back to being productive, they had implemented a temporary solution that actually deserved a thorough inspection.

Reputational Damage: The Long-Term Impact

Customers who experience late deliveries or inconsistent product quality lose confidence. Bad news spreads at lightning speed through social

media and review sites. Regaining customer trust costs far more time and money than retaining it.
A sales representative had to call three major customers to inform them that their deliveries would be delayed by one day. Two customers were understanding, but one large retail chain threatened penalties for late delivery. More concerning was the comment: "This is the second time this year. We need to talk about your reliability."

Organizational Stress: The Human Factor

Constantly putting out fires leads to frustration among personnel. Technicians become demotivated because they have to solve the same problems over and over again. Production managers face pressure from above. Good people leave the company for organizations with more reliable processes, where chaos is minimal.
Mark noticed that his best technician, Andrew, was calling in sick on Mondays more and more often. In an honest conversation, Andrew told him he was fed up with repairing the same machine over and over again. "I didn't become a technician to solve the same problem every week," he said. "I want to move forward, not keep going backward."

1.2 Reliability as a Competitive Advantage

While many companies struggle with the consequences of unreliable assets, other organizations build a structural advantage by systematically investing in reliability. These companies realize four concrete benefits that directly impact their business results.

Lower operational costs

Reliable assets run longer without interruption and have predictable maintenance costs. This makes planning easier and reduces the need for costly emergency measures. Energy efficiency remains stable, because systems operate within their design parameters.
A competitor of Mark's company had invested in a newer, more reliable packaging line. Their machine had an availability of 97% versus Mark's 89%. This meant that they could produce the same volume with less

overcapacity, lower labor costs per produced unit and much less stress in the organization.

Higher customer satisfaction

Customers value suppliers who consistently deliver what they promise, when they promise it. This leads to more repeat orders, word-of-mouth referrals, and a willingness to pay premium prices for reliable service.
Mark's competitor could guarantee customers that orders would be delivered within 48 hours — and keep that promise. This gave them a significant advantage in winning new contracts, especially with retailers who were themselves under pressure to serve their customers.

Better margins through more efficient operations

Less downtime means more production from the same assets. Stable processes reduce waste and rework. Predictable operations make better planning possible, which leads to lower inventory costs and more efficient logistics.
Through their more reliable operation, Mark's competitor was able to produce 12% more volume with the same assets, reduce inventory levels because production was predictable, and better plan maintenance during planned shutdowns and lower insurance premiums due to their low incident rates.

Stronger competitive position

You can price more aggressively while simultaneously offering better service. This makes it difficult for competitors to capture market share.
Moreover, you can invest in growth instead of constantly solving problems.
The result for Mark's competitor was impressive: 15% lower production costs and 8% higher selling prices through better service. This combination gave them an enormous lead in profitability that they could reinvest in further improvement.

1.3 The 80/20 Rule of Reliability Engineering

The most important lesson you can draw from Mark's story is that approximately 80% of all maintenance costs and reliability problems are determined during the design and construction phase of assets. This statistical truth, confirmed by industrial studies from major engineering firms and research from technical universities, has far-reaching implications for how you should approach projects.

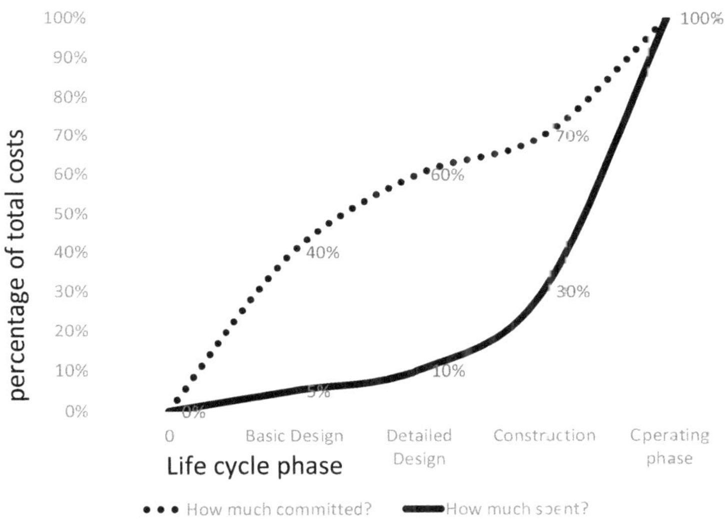

Figure 1-1 Expenditure and committed expenditure per life cycle phase

Design decisions have lifelong consequences

Industrial data from more than 500 projects in the process industry show that choosing a cheap, non-standardized component saves an average of $1,000 during construction, but causes $47,000 in additional maintenance costs over its lifetime. The decision not to build redundancy into a critical system appears to save money, but a single major failure can cost 15–20 times more than the saved redundancy.

Mark's packaging machine was originally designed with a cheaper motor that cost 15% less than the best alternative. Over three years, this "savings" of $3,000 had already led to $18,000 in additional repair and downtime costs. This ratio of 1:6 corresponds exactly with industry averages.

Quality During Construction Determines Operational Performance

Experience in industry shows time and again that even the best components can fail due to improper installation. Careless alignment of pumps causes vibrations that drastically accelerate wear. Incorrect cable installation leads to EMC problems and significantly increases the likelihood of electronics failures. On Mark's machine, the motor was installed correctly, but the alignment with the drive was not right. This misalignment of 0.3 mm caused additional vibrations that the bearings to wear significantly faster. An extra quarter hour of alignment during installation could have prevented the failure.

Training and Documentation Determine Maintenance Costs

Assets that are commissioned without proper procedures, training, and documentation perform significantly below their potential. Operators make more mistakes due to a lack of understanding of the specific requirements of their equipment. Technicians regularly fall back on standard procedures that are not suited for specific systems, or use incorrect spare parts. Mark realized that his technicians had never received thorough training on the specific motor. They treated it the same as all other motors, but this type had specific lubrication requirements they were unaware of. Research shows that improper lubrication is responsible for approximately 80% of all bearing failures — a problem that could have been entirely prevented with the right knowledge.

Maintenance Cannot Compensate for Poor Design

Years of experience in industry confirm a fundamental law: even the best maintenance can only achieve limited improvements in assets with fundamental design flaws. The vast majority of the capacity and reliability are established during design and construction. You can't make a car faster through better maintenance, and you can't cut an oven's energy consumption in half by switching to a different lubricant.

Mark's experience illustrated this well. Despite excellent maintenance by skilled technicians, the machine kept causing problems. The problem was not in the execution of the maintenance, but in the fundamental design choices that had been made years earlier. No maintenance program in the world can compensate for what has been established during the design phase.

1.4 Summary: The Core Lessons

Mark's story illustrates four fundamental truths about reliability. First, the true costs of unreliability are many times higher than the visible repair costs, due to production loss, quality problems, safety risks, reputational damage, and organizational stress.
Secondly, reliability is a competitive advantage that leads to lower operational costs, higher customer satisfaction, better margins, and a stronger market position.
80 percent of reliability problems arise during design and construction, which means early attention to reliability is far more effective than repairing after the fact.

Now that you understand why reliability is so important, the question arises: where do you begin? In the next chapter, you will learn how to move from an initiative or problem to concrete, measurable specifications — how you systematically determine exactly what you need before you start designing. For critical assets, this specification phase will be much more thorough than for standard assets, but the basic principles remain the same.

11.32

2 From Initiative to Specification

You now know why reliability is important. The next step is to move from an initiative or problem to concrete, measurable specifications. Too often, organizations jump too quickly to solutions without fully understanding the problem. This chapter shows you how to prevent that.

The Story of Tim and the Power Plant

"We need to modernize our energy infrastructure," Tim said during the monthly board meeting of Metal Precision Group, a mid-sized machine builder. "Our current systems are outdated and costing us too much money in energy and maintenance." It seemed like a straightforward observation, but no one at the table realized that this single sentence would become the starting point of a journey that would ultimately determine whether the company would struggle for the next twenty years with reliability problems or would benefit from an efficient and stable energy system.

Production manager Linda nodded in agreement. "Those old boilers and compressors do nothing but break down. Last week the entire machining department was at a standstill, because the main compressor had lost its bearings again." Quality manager Peter added: "And we haven't even talked about the variations in steam pressure that are completely throwing off our heat treatment furnaces."

CFO Hank looked up from his laptop. "What is something like this going to cost?" Tim shrugged. "No idea. Somewhere between 500,000 and 2,000,000 dollar, I'd estimate. Depends on exactly what we're going to do." Hank frowned. "That's quite a huge difference. Can't we just buy the same systems again?" Hank asked.

This conversation, which takes place daily in boardrooms throughout the country, illustrates well why so many projects ultimately disappoint. It seems as though the solution is obvious — replace the old systems with new ones — but in reality we are at the beginning of a complex journey in which crucial choices must be made that will reverberate for decades to come.

2.1 The Dangers of Drawing Conclusions Too Quickly

The greatest danger in the initiation phase is the tendency to jump straight to solutions. "We have a problem with our energy systems, so we need to buy new systems." However, this seemingly logical reasoning skips a fundamental step: thoroughly analyzing what you actually need and why.

When you dig deeper into the energy problem at Metal Precision Group, interesting aspects come to light. The current systems are indeed worn out, but why does this keep happening? Is it due to the quality of the machines or the way they are being used? Are they undersized, causing them to run continuously at full capacity? Or oversized, causing them to operate inefficiently?

Another crucial question is whether the current energy carriers are even optimal for all applications. Many companies use steam for processes that could be carried out much more efficiently with electric heating. Steam has specific advantages for certain applications — uniform temperature distribution, good heat transfer, safety at high temperatures. But for other applications, direct electric heating, heat pumps or even solar thermal energy can be far more cost-effective.

The problem with reaching conclusions too quickly is that they lead to treating symptoms rather than solving the underlying causes. You buy new boilers and compressors, but if the real problem lies in oversizing, inefficient use, or poor maintenance, you'll be facing the same problems again in a few years.

The Cost of Wrong Assumptions

Industrial research shows that projects that begin with preconceived solutions have on average 35% higher life-cycle costs than projects that start with thorough problem analysis. At Metal Precision Group, a wrong assumption could make a difference of $700,000 over the life of the system.

2.2 Systematic Problem Analysis: The 5 Whys Technique

A powerful method for getting to the root of a problem is the 5 Whys technique. This simple but effective approach helps you look beyond the symptoms to the actual causes. Let's apply this to the energy problem:

1. Why 1: Why do the energy systems need to be replaced? Because they break down frequently and consume a lot of energy.
2. Why 2: Why do they often break down? Because the boilers regularly burn out and corrosion problems occur.
3. Why 3: Why are corrosion problems occurring? Because the water treatment is not optimal and the systems often run at partial load.
4. Why 4: Why do they often run at partial load? Because energy demand varies greatly throughout the day and the current systems are not flexible enough.
5. Why 5: Why does the energy demand vary so greatly? Because different production processes require heat at different times and there is no heat recovery.

This process reveals that the real problem is not the boilers themselves, but the lack of flexibility and heat recovery. Now we can consider various solution directions: modular systems, buffer tanks, heat recovery, or even process optimization to level out the energy demand.

The 5 Whys technique works best when you assemble a diverse team that brings different perspectives. A maintenance technician sees different causes than a production manager or an energy coordinator. By combining their knowledge, you get a more complete picture of the problem.

2.3 Stakeholder Analysis: Who Has a Stake in the Solution?

Before you begin formulating solutions, it is crucial to identify all stakeholders who have an interest in the project. In the energy project at Metal Precision Group, that includes not only management and the maintenance department, but also many other parties.

The most direct users are the operators of the production lines who depend daily on stable steam and compressed air pressure. They know from experience what problems arise and how these affect their work, which makes their input indispensable for determining actual capacity requirements and operational demands.

Equally crucial is the maintenance department that is responsible for daily maintenance and troubleshooting. They know the weak points of the current system and can provide valuable input on maintainability and reliability.
Additionally, there are various departments that indirectly but significantly experience energy problems.

The logistics department experiences problems, for example, when production is halted due to energy failures, and their perspective helps in understanding the broader impact on operations. The quality department encounters deviations caused by temperature and pressure fluctuations, making their experiences crucial for establishing stability requirements.

Various supporting departments also play an important role in the project.

External parties such as energy system suppliers, installation specialists, and energy consultants also have an influence on the project. Their expertise and limitations determine which solutions are practically feasible. In some cases, neighboring residents may also be relevant if there is potential noise or emission nuisance.
The challenge lies in the fact that every stakeholder has different requirements and expectations. Where the production department primarily wants reliability and constant pressure, the finance department focuses on total lifecycle costs. The maintenance department strives for simple and safe access for maintenance, while the energy coordinator pursues maximum efficiency and sustainability.

Involving all stakeholders early prevents important requirements from being overlooked, but also requires careful coordination to keep the project from stalling due to conflicting demands. A systematic approach in which all perspectives are inventoried and weighed is therefore essential to the ultimate success of the project.

2.4 Early Equipment Management

In a classic project, it happens that a mechanic sees a machine for the first time when he is confronted with a failure on this new machine in the middle of the night. Often he has been called by an operator who is also unfamiliar with the machine. It is only human that the motivation of these people to correct the engineering mistakes is very low. They feel that the designers have no interest in the problems they are forced to deal with.

This situation arises because traditional project management was primarily the responsibility of engineering departments, while other departments were barely involved in the project. It often turns out that the greatest production losses occur with new installations. It sometimes takes months or even years before an installation reaches its design capacity and the failure rate stabilizes at a lower level.

Early Equipment Management as a Stakeholder Strategy

Early Equipment Management is a methodology of project management that focuses on reducing these losses through systematic involvement of all relevant parties from the very beginning. On the shop floor, years of knowledge have often been accumulated about the predecessors of the technical system in the new design. EEM uses this knowledge and then brings the knowledge about the new design back to that same organization.

EEM is based on the fundamental principle that the people who do the work themselves know best how it can be improved. The knowledge lies on the shop floor.

More importantly, involving production employees and the maintenance technicians in the project leads to an enthusiasm that is very important during the startup of a facility.

Systematic Multidisciplinary Collaboration

In section 2.3 we identified all relevant stakeholders for the energy project: from operators and technicians to procurement and HSE departments. Early Equipment Management provides the methodology

to engage these diverse groups systematically and effectively throughout the entire project.
The power of EEM lies in the systematic organization of multidisciplinary collaboration between three crucial streams that must run in parallel during the project.

Engineering must build, test, and ensure the updating of documents, but this now happens in close collaboration with the end users instead of in isolation.
Operations must develop Standard Operating Procedures, train its employees, and thoroughly test the installation before commissioning. Through early involvement, they can provide input on ease of operation and operational requirements during the design phase, resulting in systems that are properly operable from day one. Maintenance must develop the preventive maintenance that the installation requires, train its employees, and ensure that the supplied spare parts are properly entered into the warehouse before the installation starts. These spare parts should be organization to charge to the project budget. Through their early involvement, maintenance aspects can be incorporated into the design from the start rather than having to be addressed after the fact.

Iterative Optimization During Project Phases

A core principle of EEM is that both during the definition phase and the design phase, the teams go through an iterative process of drafting and improving, so that they achieve the most optimal possible end result. They implement improvements from the drawing stage onward, and not only when the installation has already been built. This approach prevents the common situation where problems are only discovered during commissioning.

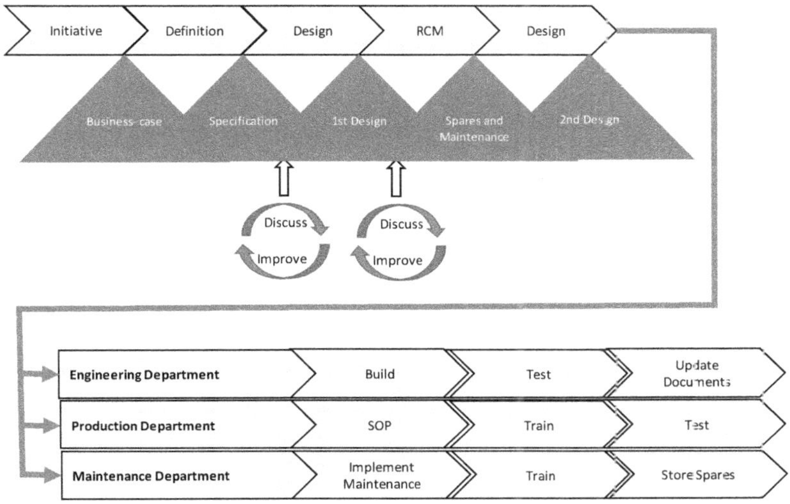

Figure 2-1 Schematic process flow of early equipment management

Some companies bring operators along when visiting machine manufacturers, so they can share their opinions and ideas about the new design right from the start. When problems arise during startup, they see it as their problem, and their attitude tends to be "let's see how we can solve this" rather than criticizing the design.

The Role of Commitment and Ownership

A crucial aspect of EEM that goes beyond traditional stakeholder management is the creation of genuine commitment and ownership. When operators and technicians have been involved in thinking through the design, they feel more responsible for the success of the system. They understand why certain choices were made and are more motivated to make the system function well.

This involvement and ownership translates directly into better performance during the startup of the installation. Teams that have participated in the design process are more inclined to view problems as challenges that can be solved, rather than as evidence that the design is flawed. This psychological aspect of EEM is often just as valuable as the technical input that is obtained.

Practical Implementation of EEM Principles

For the energy project of Metal Precision Group, an EEM approach would mean that representatives from all three main streams are present from the very first project meetings. An experienced operator can share his experiences with the current energy systems and indicate which problems occur regularly. A senior technician can point out maintenance challenges and practical solutions he has developed. A quality employee can identify critical control points where energy stability is crucial for product quality.

This input is then systematically incorporated into the specifications and design, with all stakeholders continuing to contribute ideas about solutions and improvements. The result is a system that is not only technically correct, but also practically workable from the first day of operation.

2.5 Functional versus Technical Specifications

One of the most important distinctions when drafting specifications is that between functional and technical requirements. Functional requirements describe what the system must do, while technical requirements describe how the system must achieve that.

Functional Requirements for the Energy Project

For the energy project, a functional requirement could be: "The system must deliver 2500 kg/hr of steam at 8 bar operating pressure with a maximum pressure drop of 0.2 bar during peak load." This describes what is needed without specifying how it must be achieved.
Another functional requirement: "The system must be operational within 15 minutes after a cold start." This sets a performance standard without specifying which technology must be used.

Technical Requirements for the Energy Project

A technical requirement might be: "The system must be equipped with two gas engines, each with 500 kW of thermal output." This specifies which technology must be used.

Another technical requirement: "All main steam lines must be carried out in carbon steel A106 Grade B, Schedule 40, in accordance with ASME B31.1." This specifies exactly which materials and standards must be used for the steam lines.

A third example: "The steam system must be equipped with automatic condensate drainage using thermodynamic steam traps type TD52 or equivalent." This specifies both the function and the specific technical implementation of the condensate system.

The Importance of the Right Balance

The difference is important, because functional requirements leave room for different technical solutions, while technical requirements lock in the solution. By focusing on functional requirements, you give suppliers and designers the freedom to develop creative and possibly better solutions.

In practice, you need both types of requirements. Functional requirements provide flexibility and innovation, while technical requirements ensure compatibility with existing systems and standardization. It is an art to find the right balance.

The reliability requirements are often located in a gray area between functional and technical. "The system may be down for a maximum of 16 hours per year "unplanned downtime" is functionally formulated, but has direct technical implications for redundancy, component quality, and maintenance strategy.

2.6 Quantitative Reliability Objectives

One of the most common mistakes when specifying new systems is the use of vague reliability terms. Specifications often contain phrases such as "The system must be highly reliable" or "Minimal downtime is required." These terms are completely useless, because they provide no concrete criteria for evaluating different solutions.

Concrete Reliability Requirements

Instead, reliability requirements must be quantified in measurable terms. For the energy project, concrete requirements could be:

Availability: "The system must be available at least 99.8% of the time, which amounts to a maximum of 16 hours of downtime per year."

Reliability: "The mean time between failures (MTBF) must be at least 8,760 hours for critical components." This means that the entire system of critical components may have an average of one failure per year. In other words: if you consider all critical components together, a failure in the total system may occur on average no more than once every 8,760 hours (one year).
Maintainability: "Planned maintenance may last a maximum of 8 hours at a time and must be possible without shutting down the entire production."
Performance: "Steam pressure fluctuations must not exceed ±0.3 bar during normal operation."

Quantification Based on Data

These quantitative requirements make it possible to objectively compare different suppliers and solutions. They also form the basis for contractual agreements and acceptance criteria.
Establishing realistic reliability objectives requires insight into the current situation and benchmark data from comparable systems. If your current system has 96% availability, a requirement of 99.9% may be too ambitious and unnecessarily expensive. On the other hand, if your competitors are achieving 99% availability, 96% may be too low.
Industrial benchmarks show that modern energy systems in the process industry typically achieve 99.2–99.7% availability. For discrete manufacturing, this is usually between 98.5–99.3%. These figures help in establishing realistic targets.

2.7 Class vs. Quality: Specifying Installation Quality

When drafting specifications for new assets, it is crucial to understand the difference between class and quality — two concepts that are often used interchangeably, but are fundamentally different aspects of reliability.
Quality indicates how well components meet their technical specifications. A pump that is specified to deliver 100 m^3/hour at 8 bar and actually does so has good quality.

		from BEP operating point	L/D straight suction pipe	Shaft alignment	Pipe alignment	Vibration from unbalance	Foundation design	Grouting under foundation
Good practices	Remaining life factor	+15% to -30% of BEP (*)	L/D 1 to 3	+/-0.009 mm/mm error	3 mm error	6 mm/s	0.5x equipment mass	Cement with low adhesion
Impeller	0,1987	68%	75%	90%	69%	81%	88%	88%
Pump casing	0,1466	70%	80%	83%	64%	79%	78%	80%
Pump bearings	0,0152	65%	60%	58%	40%	61%	50%	55%
Seals	0,0095	51%	60%	40%	40%	64%	55%	55%
Shafts	0,0152	65%	60%	58%	40%	61%	50%	55%
Coupling	0,1149	76%	80%	65%	71%	78%	70%	75%
Motor bearings	0,0741	78%	80%	55%	80%	75%	60%	60%
Motor windings	0,8669	97%	100%	100%	100%	95%	96%	98%
Motor rotor	0,0741	78%	80%	55%	80%	75%	60%	60%
Motor starter	1,0000	100%	100%	100%	100%	100%	100%	100%
Cost to achieve class		100%	100%	100%	100%	100%	100%	100%

		from BEP operating point	L/D straight suction pipe	Shaft alignment	Pipe alignment	Vibration from unbalance	Foundation design	Grouting under foundation
Better practices	Remaining life factor	+10% to -20% of BEP	L/D 6 to 8	+/-0.003 mm/mm error	0.25 mm error	1 mm/s	3.5x equipment mass	Slightly porous but adhesive
Impeller	0,6603	88%	95%	95%	94%	95%	95%	98%
Pump casing	0,5197	73%	95%	95%	92%	95%	95%	95%
Pump bearings	0,4014	79%	90%	88%	88%	90%	90%	90%
Seals	0,4365	88%	90%	90%	84%	90%	90%	90%
Shafts	0,4014	79%	90%	88%	88%	90%	90%	90%
Coupling	0,5753	92%	95%	90%	94%	95%	90%	91%
Motor bearings	0,6134	94%	93%	94%	97%	95%	90%	90%
Motor windings	0,9801	100%	100%	100%	100%	100%	99%	99%
Motor rotor	0,6134	94%	93%	94%	97%	95%	90%	90%
Motor starter	1,0000	100%	100%	100%	100%	100%	100%	100%
Cost to achieve class		181%	120%	120%	120%	138%	200%	200%

		from BEP operating point	L/D straight suction pipe	Shaft alignment	Pipe alignment	Vibration from unbalance	Foundation design	Grouting under foundation
Best practices	Remaining life factor	+5% to -10% of BEP	L/D 10 to 12	0.001 mm/mm error	0.1 mm error	0.5 mm/s	5x equipment mass	Solid and adhesive epoxy
Impeller	0,9800	98%	100%	100%	100%	100%	100%	100%
Pump casing	0,8600	86%	100%	100%	100%	100%	100%	100%
Pump bearings	0,9702	98%	100%	100%	100%	99%	100%	100%
Seals	0,9702	98%	99%	100%	100%	100%	100%	100%
Shafts	0,9702	98%	100%	100%	100%	99%	100%	100%
Coupling	0,9801	99%	100%	99%	100%	100%	100%	100%
Motor bearings	1,0000	100%	100%	100%	100%	100%	100%	100%
Motor windings	1,0000	100%	100%	100%	100%	100%	100%	100%
Motor rotor	1,0000	100%	100%	100%	100%	100%	100%	100%
Motor starter	1,0000	100%	100%	100%	100%	100%	100%	100%
Cost to achieve class		225%	130%	150%	150%	168%	400%	300%

Figure 2-2 Impact of class on service life

Class refers to the degree of perfection of the complete system, including the quality of components, assembly, configuration, and operational context. A high-quality pump can have a dramatically short service life if the class of the installation is low due to poor alignment, incorrect application, or inadequate foundation. Figure 2.2 shows that if a pump is aligned with an accuracy that many companies call "good," the service life of the seals will be only 40 percent of the service life the seals could achieve with the best attainable alignment.

Specifying Installation Class

Specify functional class requirements instead of technical specifications alone.
These class specifications form the basis for verification during commissioning (see Chapter 9).

2.8 Identifying Boundary Conditions and Constraints

Alongside the desired functionalities, there are always constraints and limitations that influence the design. These can be technical, economic, legal, or organizational in nature.

Technical Constraints

At the energy project, technical constraints could include: limited space in the machine hall, existing gas connection of 100 m^3/h, noise limitation to 65 dB(A) at the property boundary, and connection to existing steam and compressed air distribution.
Existing infrastructure often determines which solutions are practically feasible. A complete overhaul of distribution networks can make the project twice as expensive.

Economic Constraints

These include the maximum available budget of $1,500,000, a desired payback period of no more than 7 years and possibly a preference for financing through an operating lease.

Energy price developments have a major impact on economic feasibility. With expected gas price increases of 20%, heat recovery more economically attractive.

Legal Constraints

Compliance with PED (Pressure Equipment Directive), environmental permits for emissions and noise levels, work safety legislation for safe maintenance and possible upcoming CO2 levies must all be taken into account in the specifications.

Organizational Prerequisites

Implementation during a 6-week summer shutdown, limited availability of in-house technicians, preference for a supplier within 150 km for support, and integration with existing control systems.
Making these boundary conditions explicit prevents the discovery later in the project that certain solutions are not feasible. It also helps with selecting suppliers and evaluating quotes.

2.9 Future-Proofing and Flexibility

When specifying new systems, it is important to look not only at current needs, but also at possible future developments. Systems you install today often need to last 20–25 years, and during that period the requirements can change considerably.

Possible Future Developments

For the energy project, future developments could include: a 40% expansion of production capacity, new processes requiring different temperatures, stricter CO2 requirements from climate legislation, and integration with renewable energy sources such as solar panels.
The transition to more sustainable energy is becoming increasingly important. Systems must be prepared for hybrid solutions in which fossil energy is combined with electricity from renewable sources.

Building in Flexibility

Flexibility can be built in in several ways:

Modular design means systems that can easily be expanded. Instead of one large boiler, multiple smaller units can be installed that can be switched on and off as needed.

Overcapacity means deliberately installing 15–25% more capacity than strictly necessary. This provides room for growth and ensures better efficiency at partial load.

Provisions means reserving connections and space for future expansions. Extra cable trays, spare connections for gas and electricity, and space for additional units.

Upgrade possibilities refer to the ability to replace components with better versions. Modular control systems that can receive software updates and mechanical systems that allow component upgrades.

The art is in finding the right balance between flexibility and cost.

Infinite flexibility is unaffordable, but too little flexibility leads to premature replacement.

2.10 The Specification Document: Structure and Content

A good specification document forms the basis for successful project execution. It must be clear, complete, and unambiguous. A typical structure might be:

Project Context and Objectives

Current production loss: 15% due to energy supply failures
Energy costs are 40% above benchmark
Annual maintenance costs rising to $180,000
Objective: reliable energy supply at lower total cost

Functional Requirements

Steam capacity: 2,500 kg/h at 8 bar working pressure
Compressed air capacity: 850 m^3/h at 7 bar working pressure
Peak load: 150% for 30 minutes
Restart after power failure: operational within 20 minutes
Start-up time from cold condition: maximum 15 minutes

Reliability Requirements

Availability: 99.8% (minimum 8,742 hours per year)
MTBF critical components: minimum 8,760 hours
Planned maintenance: maximum eight hours per maintenance session
Unplanned maintenance: maximum four interruptions per year

Technical Constraints
Noise level: maximum 85 dB(A) at one meter distance
Explosion-proof design: Zone 2 classification
Existing gas connection: maximum 800 m^3/h natural gas
Available space: 15 × 20 meter installation area

Commercial Terms
Project budget: $1,500,000 (excluding VAT)
Implementation time: six months from order placement
Warranty period: minimum two years full warranty
Service: 24/7 availability with four-hour response time

Acceptance Criteria
Performance test: 72 hours continuous operation at nominal capacity
Availability test: three months monitoring without interruptions
Maintenance test: demonstration of maintenance on all critical components
Documentation: complete as-built drawings and maintenance manuals

The document must be written in such a way that different suppliers give the same interpretation to the requirements. Ambiguous formulations lead to misunderstandings and disputes.

2.11 Validation of Specifications

Before the specifications are finalized, it is wise to validate them with various parties.

Internal Validation

This means having all stakeholders review the specifications. Are all requirements complete? Are they realistic? Are there contradictions?

The review must take place systematically with sufficient time for feedback. A checklist helps to ensure nothing is overlooked.

External Validation

This involves discussing concept specifications with potential suppliers. Are the requirements technically feasible within the budget? Are there alternative approaches that might work better? This input can lead to improvements in the specifications.
Suppliers can warn against unrealistic combinations of requirements. For example: 99.9% availability at the price of a standard system is usually not achievable.

Benchmarking

This compares the requirements with similar projects and industry standards. Are the requirements realistic and competitive? Benchmarking helps establish realistic objectives.
This validation can lead to adjustments in the specifications, but prevents much larger problems later in the project.

2.12 Common Mistakes in the Specification Phase

The specification phase largely determines the success of a project, yet fundamental mistakes are regularly made that have consequences lasting for years. These mistakes can often be prevented by giving systematic attention to reliability engineering principles from the very beginning.

Premature Commitment to Technical Solutions

Instead of setting functional requirements, a specific technical solution is sometimes chosen right away. This limits innovation and can lead to suboptimal solutions. It is better to define the desired outcome to describe and give suppliers room for creativity.
An example of this is specifying "install two gas engines of 500 kW each" instead of "reliably deliver 1000 kW of electrical power with no more than four hours of unplanned downtime per year." The first approach rules out alternatives such as one larger engine with battery backup or hybrid solutions that may perform better.

Underestimating Complexity and System Reliability

A fundamental mistake is underestimating the impact of complexity on system reliability. The number of components in a system directly affects the number of failures. A system consisting of a single component has a 97% chance of having operated failure-free after one year, assuming a failure probability of 3% per component per year. However, a system consisting of 25 components has only 0.97^{25}, or a 46% chance of still operating failure-free after one year.

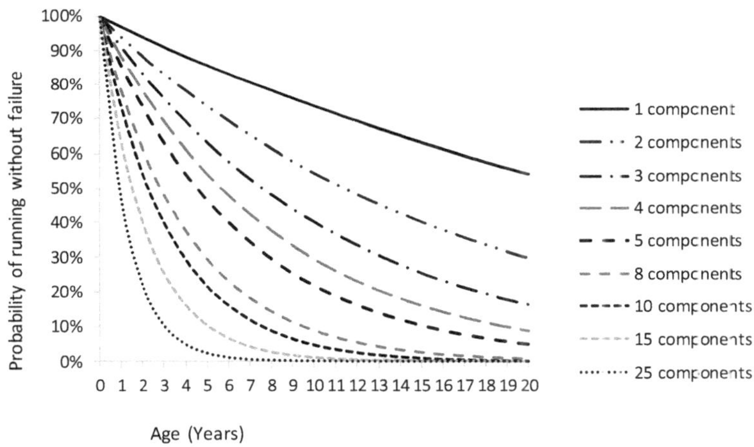

Figure 2-3 Impact of number of components on service life

As a rule of thumb, twice as many components leads to twice as many failures. This is a simplification of reality, but accurate enough for most applications. Engineers who build reliability block diagrams or develop digital twins can model this effect more precisely, but the basic principle remains: simple systems are inherently more reliable than complex systems.

This means that during specification, a deliberate assessment must be made of whether additional functionality is worth the added complexity. Every extra function, sensor, actuator, or control loop increases the probability of failures in the system. The question is not whether these components can fail, but when — and what the impact of that failure will be.

Insufficient Attention to Proven Technology

New technologies always carry a risk. Many people underestimate this. The risk is so significant that some calculation methodologies require that for technology that has been in use for less than five years, you halve the reliability in calculations. Official standards therefore assume that the probability of failure is twice as high for new technology.
The lower reliability applies not only to the component itself, but also to the entire design within a given process. New configurations, unusual applications, or innovative integrations can cause unforeseen problems that only become fully apparent after months or years of operation.
This does not mean that innovation should be avoided, but rather that conscious choices must be made. If innovative technology is chosen, this must be a well-reasoned decision with adequate risk mitigation, not an automatic preference for the newest and most modern.

Overdimensioning Through Stacked Safety Margins

A common cause of problems is oversizing through the stacking of safety margins at various stages of the design process. This problem is discussed in detail in section 7.1, but it originates during the specification phase due to a lack of communication between different disciplines. The result is equipment that operates far outside its optimal operating point, leading to reduced efficiency, shorter service life, and higher maintenance costs. Systematic checking of specifications for oversizing prevents this costly mistake.

Underestimating Interfaces

Much attention is paid to the main system, but too little to the interfaces with existing systems. This often leads to problems during implementation. Existing control systems, distribution networks, and maintenance procedures must all be included in the specifications.
Interface problems are often the most difficult to resolve, because they only become fully apparent during commissioning, when changes are

expensive and time-consuming. A systematic interface analysis during specification prevents many of these problems.

Neglecting Human Factors

Focus on technical performance, but insufficient attention to ease of use, safety, and maintainability. Systems that are technically perfect but difficult to operate or maintain often fail in practice.
Human factors encompass not only ergonomics and operation—
ease of use, but also training needs, competency requirements, and the impact of the new system on existing work processes. These aspects must already be addressed in the specification phase.

Unrealistic Reliability Requirements

Requirements that are technically unfeasible or economically unjustifiable. A requirement of 99.99% availability at standard prices is usually unrealistic. This leads to disappointment and compromises later in the project.
Realistic reliability requirements are based on benchmarking against comparable systems in the industry and a fair balance between performance and cost.

Incomplete Specifications

Important aspects are overlooked, leading to scope changes and disputes during execution. Training, documentation, and handover procedures are often underemphasized, but are essential for successful implementation. Other often overlooked aspects are spare parts strategies, long-term supplier support, methods for upgrading software and hardware, and integration requirements with future systems.

Too Little Time for Specification Development

Under pressure from project timelines, the specification phase is cut short, leading to incomplete or poorly thought-out requirements. This "savings" in the specification phase almost always results in far greater costs later in the project due to changes, adjustments, and suboptimal performance.

A thorough specification phase typically costs 5-10% of the total project time, but can prevent 30-50% of the problems during realization. The Investment in time and attention during specification pays off many times over.

Insufficient Validation of Specifications

Specifications are not verified with potential suppliers or benchmarked against comparable projects. This can lead to requirements that prove unachievable or, conversely, insufficiently ambitious.
Validation must occur systematically through internal review, external feedback, and benchmarking before specifications are finalized.

Lack of Traceability

Specifications are drawn up without a clear link to business objectives and stakeholder needs. This makes it difficult later to prioritize changes and trade-offs.
Every specification requirement must be traceable to a concrete business need or stakeholder requirement. This traceability helps in making well-informed decisions during the rest of the project.

Prevention Through a Systematic Approach

These errors are largely preventable by following the systematic approach outlined in this chapter: thorough problem analysis, stakeholder involvement, functional specifications, quantitative objectives, and validation. The extra time invested in a careful specification phase pays for itself many times over in project success and the life-cycle performance of the assets.

2.13 RAMS Methodology

RAMS - Reliability, Availability, Maintainability and Safety - is a methodology specifically developed to integrate reliability into the complete project lifecycle from the very first specification development. This approach is widely used in infrastructure, rail, and aviation, sectors where failure can have unacceptable consequences.

Systematic Integration from Specification

The power of RAMS is that it incorporates all four aspects into every decision from the very beginning of a project. When you specify a new installation, you don't just ask "Can it deliver the required capacity?" but also "How reliable does it need to be?", "How quickly can we repair it if it breaks down?" and "How do we prevent people from getting injured?"

RAMS uses a systematic approach in which every life cycle phase of a project — from concept to decommissioning — is analyzed across all four aspects. This means that maintenance strategies, spare parts, and training needs are already considered during the specification phase.

Quantitative Objectives as a Starting Point

A typical RAMS project begins with establishing quantitative objectives as discussed in section 2.13. Instead of vague terms such as "highly reliable," concrete requirements are set: "an average of no more than 2 failures per year lasting longer than 30 minutes." These concrete objectives make it possible to compare different design options and measure progress objectively.

For new projects, RAMS principles are implemented by defining reliability requirements from the start: a maximum of 4 hours of unplanned downtime per month, an average repair time of no more than 2 hours, and a minimum availability of 95% during production hours. These requirements are treated as equally important as functional requirements and influence all subsequent design decisions.

RAMS as a Framework for Further Methodologies

RAMS forms the overarching framework within which all specific reliability methodologies from the following chapters are applied. FMEA analyses (chapter 4) systematically identify how RAMS objectives can be threatened. Maintainability design (chapter 5) ensures that Maintainability objectives are realized. RCM strategies (chapter 6) support both Reliability and Availability objectives.

By using RAMS as the starting point for specification development, each subsequent chapter in this book gains a clear context and direction. Instead of standalone methodologies, they become components of a integrated approach that systematically addresses all aspects of reliability from the very first moment of a project.

2.14 Project Phases with Approval Criteria

Implementing approval criteria for reliability ensures that projects cannot proceed to the next phases without adequate attention to reliability.
Criteria after the concept phase require that all projects have quantitative reliability objectives derived from business objectives. Vague statements such as "highly reliable" are not accepted. Instead, projects must specify: "minimum 97% availability during production seasons" or "maximum two unplanned shutdowns per year lasting longer than four hours."

Criteria after the design phase evaluate whether the design of the new system can realistically meet the stated reliability objectives. This requires modeling and analysis, not just hope and intuition. Projects must demonstrate through calculations or benchmarking that their designs will deliver the required performance. Criteria regarding the construction phase ensure that contracts with suppliers contain adequate performance guarantees and that implementation plans give sufficient attention to commissioning, testing, and training. Projects that treat these aspects as "details for later" will not receive approval for realization.
Criteria during commissioning require that systems truly meet the promised reliability requirements before being considered successful. This often means longer acceptance test periods and more thorough performance monitoring than is traditionally customary.

2.15 Resource Planning for Reliability Engineering

Effective reliability engineering requires dedicated, competent resources who have sufficient time to deliver thorough work.
Managers must realize that reliability engineering requires specialized skills that are not present in everyone. Identify who in the organization has experience with FMEA, life cycle cost calculations, or reliability modeling, and ensure that these people are available for critical projects.
For projects where internal expertise is lacking, plan for external consultants or training programs. It is cheaper to train people properly or to

temporarily bring in expertise than to make costly mistakes due to insufficient analysis.

Budget for reliability engineering must be explicitly allocated rather than hidden within general engineering costs. Projects must reserve specific budget for FMEA studies, reliability testing, extended commissioning periods, and training programs. This makes the investment visible, but also prevents reliability activities from being cut under budget pressure.

Thorough reliability engineering takes time. Projects that do not allow for this time scheduling, ending up with superficial analyses that add little value. Better results are achieved by doing fewer projects thoroughly than by handling many projects superficially.

2.16 Summary

The phase from initiative to specification is crucial for the ultimate success of a project. It determines not only what is built, but also how well the system will function throughout its lifespan. Careful problem analysis, stakeholder involvement, and quantitative specifications are essential for success.

The energy project at Metal Precision Group illustrates how a seemingly simple replacement project can grow into a strategic decision about energy use, productivity, and reliability. By working systematically and thinking through all aspects, a project can be transformed from treating symptoms to achieving structural improvement.

The most important lessons are clear. Use the 5 Whys technique to get to the root causes. Involve all stakeholders early in the process. Formulate requirements functionally rather than technically where possible. Quantify reliability objectives with concrete, measurable criteria. Validate specifications thoroughly before moving forward.

Not all assets deserve the same attention. In the next chapter, you will learn how criticality analysis helps you systematically set priorities and focus your resources on the assets where reliability makes the most difference.

3 Criticality Analysis

You saw in Chapter 1 why reliability is so valuable. The reality is that 80% of maintenance costs and reliability is determined during design. But where do you start? Which assets deserve extra attention during engineering?

Not all assets are equal. Some are critical for safety, others for production continuity, and still others for quality or the environment. The failure of a pump in a bypass line may be an inconvenience, but the failure of a pump in the main stream can shut down the entire plant. This chapter shows you how to systematically set priorities using criticality analysis. This is not just a convenient tool — it is the most important decision you make in the early phase of a project. This analysis determines whether you spend $5,000 or $50,000 in engineering per asset, whether you buy standard assets or create custom designs, and whether your operations run reliably or struggle with problems for years.

You learn to assess assets on five risk dimensions and translate these scores into concrete design decisions. This methodology forms the foundation for all subsequent steps in reliability engineering — it is the first analytical step that determines where your attention should go.

3.1 When and Why

In a typical engineering project, you go through several phases: concept, basic design, detail engineering, procurement, construction, and commissioning. Most companies don't start thinking about reliability until detail engineering, or even as late as commissioning. By then, it's too late.
The crucial moment for criticality analysis is as soon as your basic design is complete. At that point, you know which assets will be in your system — which machines your production line contains or which equipment your process flow diagram shows — but not yet exactly how they will be designed. This is precisely the right moment.
Why is timing so important? Because this analysis shapes everything that follows, determines. For a critical asset, as a company you need to actively

participate in the design process, perform extensive FMEA and RCM analyses, develop custom maintenance programs, and conduct careful supplier selection. For a standard asset, it is sufficient to choose a reliable brand from a good supplier.

3.2 Sarah's Story: Three Days That Changed Everything

Sarah was the maintenance manager at TechPro Manufacturing, a mid-sized producer of precision components. The company was about to invest $4.5 million in a new production line with approximately 200 assets. In the past, TechPro's engineering department had treated all equipment more or less equally during the design process — every machine received a little attention, no machine received much attention.

"We had learned from our mistakes," Sarah said. "Our previous production line struggled with structural problems. The main compressor failed monthly because nobody had thought about the 24/7 load and dusty environment. At the same time, we had put three months of engineering into the office equipment, which was completely excessive."

This time Sarah wanted to approach it differently. As soon as the basic design was ready and she knew which 200 assets the new line would contain, she called her team together in the conference room: maintenance supervisor John, process engineer Lisa, production leader Mark, and operator Andrew.

"We're going to systematically assess all 200 assets," she began. "3 sessions of 3 hours each over 3 days, but then we'll finally know which machines are truly critical."

James had his doubts. "3 sessions for 200 machines? That's less than 3 minutes per machine. Can that really be thorough enough?"

"We are not going to design the maintenance," Sarah explained. "We are only determining which assets deserve a lot of attention and which deserve little. For that thorough analysis, we will then have all the time we need for the assets that truly justify it."

After 3 intensive sessions of 3 hours each, they had a clear picture: 25 assets were critical, 75 important, and 100 standard.

"Now the difference becomes clear," James said, looking at the results. "Not all assets are equal. The main compressor with a score of 5 on economics gets 120 hours of engineering attention. The office printer with a score of 1 across the board gets 15 minutes."

3.3 The Risk Matrix

Before you can assess assets, you first need to know what acceptable risks are for your organization. This is documented in a risk matrix that is established by senior management. This matrix defines, for different risk types and different severity levels, how often an incident may occur at most.

The risk matrix answers questions such as: do we accept a safety incident with lost time occurring once every three years at our company? How many environmental incidents with local impact do we tolerate? What degree of production downtime is acceptable? This matrix forms the basis for all further risk assessment and helps with interpreting the scores you assign during criticality analysis.

Crucial warning: The risk matrix indicates acceptable frequencies for your entire organization or plant, not for an individual machine. If management considers one accident per three years acceptable for the entire plant, this does not mean that every machine is allowed to cause one accident every three years.

The matrix is fundamentally different from the criticality analysis itself. The risk matrix comes from management – they determine what the organization is willing to tolerate. The criticality analysis is your work – you determine which individual assets can cause those risks.

3.4 Types of Risks

For most organizations, five core risks are relevant. The exact definition of each score differs by company and must be determined by management. Below you will find examples that you can adapt to your own situation.
Safety impact evaluates the potential consequences for people. A score of 1 means no safety risk upon failure. A score of 2 indicates possible minor

injury that can be treated with first aid. A score of 3 means possible personnel injury requiring medical treatment. A score of 4 represents serious injury with permanent consequences. A score of 5 indicates potentially fatal accidents.

Figure 3-1 The consequences of a safety incident can be dramatic

Environmental impact assesses possible contamination or ecological damage. A score of 1 means no environmental effect. A score of 2 indicates very limited, immediately recoverable impact within the site. A score of 3 means limited local impact recoverable within days. A score of 4 represents significant off-site impact with a recovery time of weeks. A score of 5 means severe environmental contamination with long-term or permanent consequences.

Delivery reliability measures the consequences for operations. A score of 1 means no or very brief downtime (less than 30 minutes) without a workaround. A score of 2 indicates several hours of downtime or reduced capacity. Score 3 means one day of downtime or significant capacity loss. Score 4 represents multiple days of downtime with possible loss of customer orders. Score 5 means very long downtime leading to permanent customer loss.

Quality impact evaluates the consequences for product quality. A score of 1 means no quality problem. A score of 2 indicates temporary internal quality variation without customer impact. A score of 3 means quality issues that can be resolved internally before delivery. A score of 4 represents quality problems that reach customers and cause returns. A score of 5 means serious product quality problems or recalls.

Sustainability impact assesses effects on energy efficiency, CO_2 emissions, and raw material consumption. A score of 1 means no effect on environmental objectives. A score of 2 indicates a slight temporary increase in energy consumption or emissions. A score of 3 means a moderate increase in energy consumption or material waste. A score of 4 represents a considerable negative impact on annual objectives. A score of 5 means significant negative impact whereby legal limits or stakeholder commitments are jeopardized.

As I noted earlier: the above definitions for each score are merely examples. Your company's management must determine the scores. For this purpose, the risk matrix, figure 3.2, is a useful tool.
The matrix shows the definitions of the scores for the severity of a risk and how often it may occur.

Score	Economics	Safety	Condition	Energy consump-tion	Environ-ment	Heard of it	<1x per 10 years	<1x per year	<1x per month	>1x per month
1	< $1,100	No impact	No impact	Optimal	No conse-quences					
2	< $5,500	First aid	Minor complaint	< 5% increase	Internal spill					
3	< $55,000	Lost-time accident	Return delivery to customer	< 10% increase	Reportable incident					
4	< $550,000	Permanent disability		> 10% increase	Off-site damage					
5	> $550,000	Fatality	Product recall		Permanent off-site damage					

Risks in this category require immediate action

Risks in this category require planned action. These must be weighed against each other and prioritized

Risks in this category are acceptable

Figure 3-2 Example of a risk matrix

Not all scores need to have a definition — some fi4elds may remain blank. A management team can decide that certain safety risks are always more serious than the most significant delivery reliability issue. Each directorate must determine for its own organization what its requirements are and establish which scores are relevant.

Depending on your sector, other risk types may also be relevant, such as reputational damage, cybersecurity, employee satisfaction, or industry-specific risks such as food safety or data security.

3.5 The Criticality Ranking: How You Do It

Now that you know which criteria management considers important, you can begin systematically assessing assets. This process is called criticality ranking and takes place in team sessions with carefully selected participants.

Team Composition: Who and Why

A criticality ranking team typically consists of three to seven people who bring different perspectives. A larger group often leads to less efficient discussions without substantial added value. A typical composition includes: A maintenance engineer brings knowledge of asset design and maintenance techniques. He knows how machines fail and which design choices influence reliability.

A maintenance supervisor or experienced technician has practical knowledge of what really happens in the field. He knows the problems that occur and often has an intuitive sense of which assets are problematic.

A process engineer understands process requirements and knows how failures propagate through the entire production process. He can assess whether workarounds are available when a specific asset fails.

A production manager has insight into operational impact and knows how failures affect daily production. He understands the relationship with customers and can assess which disruptions are acceptable.

An experienced operator works with the equipment daily and has the best feel for which assets are truly reliable and which ones constantly cause problems.

Sarah's team at TechPro was optimally assembled. "John knew every machine inside and out," she said. "Lisa understood exactly how failures rippled through the process. Mark knew which machines we could afford to lose and which ones we couldn't. And Andrew, our operator, had the best feel for which machines were truly problematic."

The Facilitator: Guardian of the Process

You have the sessions facilitated by a facilitator who is independent of the other team members. This must be someone who knows the methodology and ensures that the team stays focused on the task.

Sometimes people want to go too deep into all kinds of possible scenarios, or a discussion arises about a previously occurred incident. The risk then exists that a discussion emerges about who is to blame, which gets in the way of open conversation. In other cases, team members have insufficient information to fill in a well-substantiated score. The facilitator must assess this, ask follow-up questions if necessary, and may decide that the team first needs to look up more information before filling in the score for a particular asset.

The Session Itself: Systematic and Structured

You conduct the criticality ranking in a quiet room where the team systematically works through the entire asset list. A session typically lasts 2 to 3 hours. Shorter sessions give people the feeling that they have to stop just when they're hitting their stride. Longer sessions lead to loss of concentration — after 3 hours, the quality of decisions declines.
For each asset, the team determines the potential impact of a realistic worst-case scenario on the various aspects of the risk matrix. The focus is not on theoretically conceivable disasters, but on what can reasonably happen when this specific asset fails.

The team discusses each asset briefly but thoroughly. Someone names the asset, another immediately gives their initial assessment, others add to it or correct it. If the team quickly reaches consensus, you fill in the scores and move on to the next one. If discussion arises, the facilitator must guide it to a conclusion or decide that the team first needs more information.

Scoring Table: Concrete Examples

For a production line, the scores might look like this, for example:

Asset	Safety	Environment	Economic	Condition	Delivery Reliability	Priority
Heat exchanger HEX-58	3	2	5	4	5	Critical
Cooling water system	2	3	4	3	4	Critical
Main compressor	3	2	5	4	5	Critical
Packaging machine	1	1	3	4	3	Critical
Conveyor A12	2	1	2	2	2	Important
Hall lighting	2	1	1	1	1	Important
Office printer	1	1	1	1	1	Standard

Figure 3-3 Score of criticality ranking

Heat exchanger HEX-58 receives a score of 3 for safety (hot water can cause injury), 2 for environment (limited glycol leakage possible), 5 for economic impact ($3,500 per hour downtime), 4 for quality (temperature variation directly affects the product), and 5 for delivery reliability (main process stops completely for two weeks in a worst case scenario). This results in the priority classification critical.

The packaging machine receives a score of 1 for safety, 1 for environment, 3 for economic impact, but 4 for quality (packaging defects can lead to product recalls) and 3 for delivery reliability. This results in a critical priority rating due to the high quality impact, despite lower scores on other dimensions.

The office printer scores 1 on all aspects, which results in the priority standard.
Sarah's team methodically scored TechPro's 200 assets according to this process. "The main compressor received a score of 5 economically — failure would cost $3,500 per hour," Sarah explained. "The office printer scored 1 across the board. The difference was immediately clear: 25 critical, 75 important, 100 standard."

3.6 From Score to Strategy: The Consequences

Now comes the crucial part: what do you do with these scores? The answer determines the success of your entire project. The criticality score translates directly into fundamentally different engineering approaches.

Three Categories, Three Totally Different Worlds

Based on the scores, you divide assets into three categories. This classification is not an administrative exercise — they are three completely different universes in terms of engineering investment, vendor selection, and operational approach.

Critical Assets: Maximum Engineering Involvement

Assets that score 4 or 5 on one or more aspects are considered critical. For these assets, a fundamentally different approach applies than for the rest. Engineering involvement from the start: Your company actively participates in the design process. This does not mean that you design the assets yourself, but that your maintenance engineers, process engineers, and operational personnel are involved from the specification phase onward. You question the supplier on design choices, follow up on material selections, and require insight into the reasoning behind the design.

No 'off-the-shelf' solutions: You don't purchase a standard asset without thorough analysis. Even if the supplier says "this is our standard model running at a hundred other customers' sites," you need to be critical. Those other customers probably have different operational conditions than you do.
Performing FMEA: You systematically perform a Failure Mode and Effect Analysis for this asset (you will learn this in chapter 4). This allows you to identify all possible failure modes, their causes, and effects. This helps you make design choices that prevent failure or limit its impact.
RCM application: You develop a customized maintenance strategy with Reliability Centered Maintenance (see chapter 6). This does not mean that you simply adopt the manufacturer's standard maintenance schedule. You systematically analyze which maintenance is truly necessary for your specific operational conditions.

Maintenance tailored to your situation: The maintenance plan you develop must fit your circumstances perfectly. This includes:
Run hours: A compressor that runs 2,000 hours per year requires different maintenance than the same compressor running 8,000 hours. The manufacturer's guideline is based on their assumptions, not your reality.

Environmental conditions: A pump in a clean, climate-controlled room ages differently than the same pump in a dusty, humid outdoor environment with temperature fluctuations from -10°C to +35°C.
Load: A motor running continuously at 90% capacity has different maintenance than the same motor that usually runs at 40% with occasional peaks.
Start-stop cycles: Equipment that runs continuously ages differently than equipment that is started and stopped daily. Thermal cycles and mechanical stress during startup are often the greatest reliability threats.

Spare parts inventory: For critical assets, you stock parts whose failure would lead to extended downtime. This often means $15,000 to $50,000 in investment per critical asset, but it prevents you from waiting weeks for a special part while your production is at a standstill.

Supplier Selection: You conduct a thorough analysis of suppliers: support organization, response times, geographic presence, and reputation. This goes far beyond price and standard specifications. You visit reference sites, speak with other users, and require detailed service level agreements (see Chapter 10).

The engineering investment for a critical asset is much higher than for a standard asset. But it is worth it — one unexpected shutdown of a critical asset can recoup this investment in a single stroke.

> Concrete example of Sarah's main compressor: "For the main compressor with economic score 5, our team spent 120 hours on engineering. We carried out an extensive FMEA and developed a maintenance plan specifically for our 24/7 operation with 8,000 running hours per year in a dusty environment. We stocked $15,000 in spare parts, including a complete set of bearings, seals, and critical sensors. We selected a supplier with a guaranteed 4-hour response time and local presence. Our compressor has now been running for 18 months without unplanned downtime. On our previous line, the compressor cost us an average of $42,000 per year in failures."

Important Assets: Pragmatic Middle Ground

Assets that score a 2 or 3 on one or more aspects, are considered important. These deserve attention, but not the same intensity as critical assets.

A reliability engineer analyzes these assets independently, not in a full team setting as with critical assets. This saves time and resources, while still ensuring systematic analysis. The engineer uses simplified versions of FMEA and RCM, such as Industrial RCM (iRCM).

Results reviewed in a team setting: Although the analysis is done solo, discusses his results with colleagues from maintenance and production. This ensures that practice is tested and that the results accepted by those who have to work with it.

Pragmatic approach: You don't perform a full classical RCM, but you do conduct a systematic analysis of the most important failure modes and maintenance tasks. This is the core of Industrial RCM - 20% of the time for 80% of the result.

Standard spare parts: You stock only the most common fast-wearing parts. For larger components, you rely on normal supplier lead times, with an express delivery option if needed.

Supplier: You choose reputable suppliers, but do not conduct an extensive selection procedure as you would for critical assets. The price-quality ratio is the deciding factor.

This is sufficient for a thorough analysis without the intensity of critical assets.

Standard Assets: Efficient and Pragmatic

Assets that score 1 on all aspects are considered standard. These assets receive minimal engineering attention.

Engineering chooses independently: The engineering department selects suitable assets without extensive team analysis or methodologies. Their knowledge and experience are sufficient.

Good brands, reliable suppliers: You buy from well-known brands through reliable suppliers. This is sufficient for this category. Price is an important factor, but you do not choose unknown suppliers purely on the basis of lowest price.

Standard maintenance: You follow the maintenance recommendations of the manufacturer or general best practices. No custom analyses needed.

Focus is on the critical and important assets.

Concrete example of Sarah's office printer: "The office printer scored 1 across the board. Our purchasing department ordered an HP LaserJet from the corporate supplier we were already using. Total engineering time: 15 minutes to check the specification. No FMEA, no RCM, no special supplier selection. If the printer breaks down, we order a new one. The downtime costs us maybe $50, the printer costs $400 — it makes no sense to spend engineering time on that."

3.7 The Business Case Impact: Concrete Numbers

The difference in approach between the three categories has dramatic impact on your project budget and resources. Let's take Sarah's project as an example with 200 assets.

Scenario A: Everything gets the same attention (the old TechPro)

- 200 assets × 80 hours average = 16,000 engineering hours
- 16,000 hours × $100/hour = $1,600,000 engineering costs

Scenario B: Differentiated Approach Based on Criticality (the new TechPro)

- 25 critical assets × 80 hours = 2,000 hours
- 75 important assets × 8 hours = 600 hours
- 100 standard assets × 0.5 hours = 50 hours
- Total = 2,650 engineering hours

2,650 hours × $100/hour = $265,000 engineering costs
Savings: $1,335,000 on engineering alone

But the story doesn't end there. In Scenario A, you're not only wasting money on over-engineering standard assets, you're also under-investing in the critical assets. 80 hours is completely insufficient for a truly critical asset. The result is that your critical assets underperform.
In Scenario B, you invest the saved resources precisely in the assets that justify it. The 25 critical assets now each receive an average of 80 hours of attention (some even 120 hours), which is sufficient for thorough analysis.

The result: 18% higher availability, $250,000 per year in reduced failure costs, and satisfied operators who can work with reliable equipment.

Legal Obligations

Alongside the economic justification, there is another fundamental reason for criticality assessment: regulatory requirements. Various laws and Regulations require that organizations know which assets are critical.

For critical assets, you must conduct a thorough analysis to identify possible failure modes and develop preventive measures. Companies must be able to demonstrate that they have performed this maintenance in a timely and correct manner. During inspections or audits, you must be able to show documentation that demonstrates you have systematically determined what is critical and that your maintenance plan gives adequate attention to these assets.

This legal obligation means that criticality assessment is not only a practical tool for efficient resource allocation, but also a compliance requirement. You must document how you determined the criticality and be able to show the justification.

3.8 Efficiency and Pragmatism

One of the most compelling arguments for criticality ranking is the speed at which you achieve results. This is not just a little faster — it is an order of magnitude difference.

A typical criticality ranking session lasts approximately 3 hours, during which a team can assess several hundred assets. Sarah's team did 200 assets in 3 days (9 hours total). This means approximately 2.7 minutes per asset to make a decision that has hundreds of hours of engineering time and thousands of dollars in impact.

Compare this with alternative methodologies. If you were to use Failure Mode and Effect Analysis (FMEA) to determine which assets are critical, a team could easily need a year. A thorough FMEA of a complex asset takes 8 to 16 hours. For 200 assets, this means 1,600 to 3,200 hours of team time — just to determine what is important, before you even begin developing maintenance strategies.

During this analysis phase you perform no concrete maintenance actions, which means that potential improvements can only be realized much later. The criticality ranking offers a pragmatic balance between thoroughness and efficiency. You invest three days to determine what you will be working on for the next three months.
This speed has another advantage: you can repeat the assessment regularly. If your business conditions change, you have new assets installs new equipment or gains new insights into the impact of failures, you can update the criticality ranking without a major time investment. Sarah's team now schedules an annual half-day update session to add new assets and revise scores for assets where experience has proven different from what was expected.

3.9 Quality Assurance of the Process

To ensure the quality of your criticality assessment, good documentation essential. This serves two purposes: it makes the process repeatable and traceable, and it meets legal documentation requirements.
Record which criteria you have used—the exact definitions of score 1 through 5 for each risk dimension. These definitions can change over time as your organization evolves or as new insights emerge. Without documentation, a year later you no longer know why the scores were assigned.
Document who was on the team and why these individuals were selected. This helps if questions arise later about the validity of scores. It also shows that you had a multidisciplinary team with relevant expertise.

Document the reasoning behind scores that sparked discussion or where the team was uncertain. If an asset could receive a score of 3 or 4, for example, and the team chose 4, note the reason why. This helps during later reviews and prevents the same discussion from having to be repeated.
It is also valuable to validate the results by checking whether the outcomes align with your intuitive sense of which assets are critical. If the methodology identifies assets as critical that nobody expected, this can mean one of two things: either the methodology has uncovered a hidden

risk that is important, or the scoring is incorrect and you need to revise the criteria.

The same applies in reverse: if assets that everyone considers critical receive low scores, you need to critically examine your criteria. Are you missing a relevant risk dimension? Are the definitions per score not sharp enough? Did the team consider the wrong worst-case scenarios?

For example, Sarah's team discovered that their lighting initially received low scores. Upon closer examination, they realized that lighting failure during the night shift posed serious safety risks. They adjusted the safety score to 2, which moved the asset into the 'important' category and ensured that attention was paid to backup systems.
Finally, it is wise to review the assessment after 6 to 12 months of operational experience. If assets turn out to fail more frequently than expected, or prove much more reliable, you can adjust scores based on this experience. The assessment is a living document, not a fixed judgment.

3.10 Summary: The Core Lessons

Criticality analysis is the most important decision you make in the early stages of an engineering project. Three core principles are central.
First, timing is crucial. Conduct the criticality analysis as soon as you basic design is complete - once you know which assets will be in your system, but before detailed engineering begins. This analysis determines how you allocate your engineering resources the coming months or years. Analyzing too late means missed opportunities and wasted resources. Analyzing too early means you don't yet know what you need to assess.
Second, different criticality levels require fundamentally different approaches. Critical assets receive full engineering involvement: actively contributing during design, performing FMEA and RCM, maintenance tailored to your operational conditions, spare parts inventory, and careful supplier selection. Important assets receive pragmatic solo analysis by a reliability engineer, with results validated in a team setting. Standard assets are adequately served by quality brands from reliable suppliers, selected by engineering without extensive analysis.

Third, systematic prioritization leads to dramatic improvements in your efficiency. By investing three days in criticality ranking, you save hundreds of hours of wasted engineering time on unimportant assets, while at the same time you can give much more attention to the assets that are truly critical. Sarah's project saved $1.3 million in engineering costs while availability rose 18% - simply by focusing resources where they make the most difference.

The criticality ranking you have now completed forms the basis for all further decisions about your project.

In the next chapter, we will learn how FMEA (Failure Mode and Effect Analysis) helps you systematically analyze what can go wrong with your system and how you can prevent it. Because the best specifications in the world are useless if they are not translated into a design that is truly reliable.

4 FMEA – Systematic Failure Analysis

The Story of Design Engineer Marie

Mary stared at her computer screen and felt her stomach turn. The email from the quality manager at Tromp Bakery was short but harsh: "Pneumatic gripper defective - production line completely down - third time this month - this cannot go on any longer."
As a senior design engineer at TechnoSystems, Mary had designed the new packaging machine for Tromp Bakery. On paper, it was a solid technical design: 120 loaves per minute, precision down to the millimeter, and fully automated. All Factory Acceptance Tests had been passed. The machine did exactly what the specification required.

But now, three months after installation, it became clear that functioning and functioning reliably are two completely different things. The pneumatic gripper that picked up the loaves and placed them in the packaging worked perfectly under laboratory conditions. In the dusty, warm bakery environment, however, bread crumbs regularly clogged the gripper's fine air channels. The result: reduced suction force, falling loaves, and complete production shutdowns.
"We focused on what the machine had to do," Mary realized.
Marie thought to herself as she read through the failure reports. "But we have never systematically analyzed what could go wrong." This experience would become the starting point for a different design approach in which reliability was not something that was considered only after the fact, but the heart of the design process.

4.1 FMEA: The Core of Reliable Design

Failure Mode and Effect Analysis (FMEA) is the most powerful methodology for systematically analyzing potential failure mechanisms during the design phase. Where traditional design focuses on achieving desired functionality, FMEA forces you to look further: what can go wrong and how can we prevent it?

The difference between functional and reliable design became clear in Marie's packaging machine. From a functional standpoint, the pneumatic gripper was perfect: it could pick up, move, and place loaves of bread at the required speed and precision. All tests under controlled conditions had passed. But no one had systematically analyzed how the gripper would perform in the actual operational environment of a commercial bakery.

FMEA could have prevented this problem by asking a simple question: "In what ways can this gripper fail to perform its function?" By systematically answering this question for each component, the team would have discovered that small particles can block the air channels could become clogged, that the gripper had not been designed for dusty environments, and that no filtering or self-cleaning mechanisms had been built in.

The Industrial Foundation of FMEA

FMEA was originally developed in the aerospace industry, where failure can literally be a matter of life or death. It has since grown into the global standard for reliability analysis across all industries. Research by automotive manufacturers shows that systematic FMEA implementation can reduce design-related failures by 60-80%.

The strength of FMEA lies in its systematic, team-oriented approach. It is not an individual brainstorming session, but a structured process in which different areas of expertise are combined to identify all possible failure modes.

4.2 The Seven Core Questions of FMEA

FMEA begins with breaking down the design into manageable components and subsystems. For each component, seven questions are then asked that form the basis of the analysis:

Question 1: What is the function of this component? For Marie's pneumatic gripper, the function would be: "Pick up loaves and transfer them from the conveyor belt to the packaging without damage, with a cycle time of no more than 3 seconds." This functional definition must be specific and measurable.

Question 2: In what ways can this component fail to perform its function? Here you identify all possible failure modes: the gripper does not pick up the loaves, it drops the loaves, it does not transport the loaves, it transports the loaves to the wrong location, it damages the loaves, the cycle time is longer than three seconds. Think broadly — from complete failure to partial performance degradation.

'**Question 3:** What are the possible causes of each failure mode?' For failure to pick up the loaves: loss of suction with the following underlying causes: clogged air channels due to crumbs, leakage in pneumatic connections, defective vacuum pump, insufficient air pressure in the system, or wear of seals. Continue down to the root causes, not just the direct symptoms.

'**Question 4:** What are the consequences if this failure mode occurs?' Loss of suction force leads to loaves falling and getting damaged, incomplete packages that do not pass quality control, production downtime while the problem is being resolved, potential food safety issues due to fallen products, and reputational damage with the customer.

Question 5: How often is this failure mode likely to occur? Rate the probability of failure on a scale of 1 to 10, where 1 means very unlikely and 10 means almost certain. For clogged air channels in a dusty bakery environment, for example, this could be a 7.

Question 6: How severe are the consequences of this failure mode? Assess the impact on a scale of 1 to 10, where 1 means minimal impact and 10 means severe consequences. A complete production shutdown in a commercial bakery, for example, might receive an 8 due to the high costs of downtime.

Question 7: How easily can this failure mode be detected before it leads to problems? Rate the detectability on a scale of one to ten, where one means the problem is immediately noticeable and ten means it is impossible to detect before failure occurs.

Gradual loss of suction power due to blockages may be difficult to detect until it is too late, so this could be a seven.

Risk Priority Number (RPN) Calculation and Prioritization

The Risk Priority Number (RPN) is the heart of the FMEA methodology and is calculated by multiplying the scores for occurrence, severity, and detectability:

$$\text{RPN} = \text{Probability} \times \text{Severity} \times \text{Detectability}$$

For Marie's gripper problem with clogged air channels, the calculation would be:

$$7\ (\text{probability}) \times 8\ (\text{severity}) \times 7\ (\text{detectability}) = 392$$

This RPN of 392 places this problem in the "high priority" category that requires immediate attention.

RPN Threshold Values and Prioritization

The RPN can range from 1 (very low risk) to 1,000 (very high risk). In practice, organizations often use the following guidelines:

- RPN > 150: Immediate action required within 2 weeks
- RPN 100–150: Plan actions within 3 months
- RPN < 100: Possibly acceptable, monitor and evaluate

Industrial data from more than 1,000 FMEA studies shows that these threshold values work well for balancing risk management and resource efficiency.

The RPN as an Improvement Tool

The RPN is more than just a prioritization tool. It also shows which risk factors you can influence to lower the overall risk score.
You can reduce the probability by improving the design or using higher-quality components. For the gripper, this could mean: implementing self-cleaning filter systems, using larger air channels that clog less easily, or applying a different gripping mechanism that is less susceptible to contamination.

You can reduce the severity by limiting the impact of failure. This could mean: installing redundant grippers so that production can continue if one gripper fails, implementing a backup system that allows manual operation, or designing a graceful degradation where the machine can continue running at a lower speed.

You can improve detectability by adding sensors or monitoring systems. For the gripper, this could mean: pressure sensors that detect decreasing suction force, flow meters that signal blockages in air channels, or timer systems that detect when gripping takes abnormally long.

What's interesting about RPN analysis is that often one of these three factors makes the dominant contribution to the high risk. In Mary's case, poor detectability was a major problem — blockages were only noticed when loaves started falling. By adding simple pressure sensors that would detect decreasing suction force, the detectability could drop from 7 to 3, reducing the RPN from 392 to 168 — still requiring attention, but much more manageable.

4.3 FMEA Team Composition and Process

The quality of an FMEA analysis depends strongly on the composition of the analysis team. A good FMEA is never the work of one person, but of a multidisciplinary team that brings different perspectives and areas of expertise to the table.

The Ideal FMEA Team

For Marie's packaging machine, an ideal FMEA team would have consisted of:

- The design engineer (Marie herself) who designed the system and knows the technical details. She can explain how the system is supposed to function, what design choices were made, and what technical limitations exist.
- A maintenance engineer or experienced technician who knows what typically goes wrong with similar machines. This person brings practical experience regarding common failures, wear patterns, and maintenance challenges.

- A production specialist who understands how the system is used and what the consequences of failures are for production. This person can assess the severity of different failure modes from an operational perspective.
- A quality specialist who knows how failure can lead to product quality problems and what the consequences are for customers. In the food industry, this is particularly critical, because quality problems can lead to food safety issues.

FMEA Time Estimates

A thorough FMEA for a critical subsystem typically requires 8-16 hours of team time, spread across 2-3 sessions. For complex systems such as Marie's complete packaging machine, the total FMEA would cost 40-60 hours of team time, but this prevents an average of 200-400 hours of problem-solving during the operational phase.

The FMEA Process

The FMEA process itself follows a structured approach:

Preparation involves assembling the team, collecting relevant documentation (drawings, specifications, comparable systems), and defining the scope of the analysis.

Function identification in which the intended function of each component is clearly defined. This may seem obvious, but it often turns out that different team members have different views on exactly what a component is supposed to do.

Failure mode identification in which all the ways in which each component can fail are systematically identified in which each component can fail. It is important to think broadly: not only complete failure, but also partial failure, intermittent problems, or performance degradation.

Cause analysis in which the underlying causes are identified for each failure mode. This is often the most valuable part of the analysis, because it directly leads to preventive measures.

Consequence analysis that determines what happens when the failure mode occurs. This includes not only direct technical consequences, but also operational, economic, safety, and quality consequences.

Risk assessment where the scores for probability, severity, and detectability are assigned and the RPN is calculated.

Action planning in which improvement actions are defined for all high-risk items (RPN > 150), responsibilities are assigned, and timelines are established.

4.4 From FMEA to Design Improvements

The goal of FMEA is not to create lists of potential problems, but to actually improve the design to prevent those problems. Mary's experience perfectly illustrates how FMEA insights can be translated into concrete design improvements.

Reliability Design Principles

After her experience with the failure-prone gripper, Mary went back to the FMEA analysis that her team had conducted after the fact. The analysis had correctly identified the clogging problem with an RPN of 392. But the FMEA had also suggested several design principles that could have prevented this problem.

Simplicity is an important design principle for reliability. Simple designs have fewer components that can fail and are often inherently more reliable than complex alternatives. For the gripper, this could have meant choosing a mechanical gripper instead of a pneumatic one.

Redundancy means building in backup systems for critical functions. For the packaging machine, this could have meant installing two grippers that work alternately, so that if one gripper fails, production can continue with the other.

Fail-safe design ensures that systems transition to a safe state upon failure. For the gripper, this could mean that upon loss of air pressure, the gripper automatically moves to the "open" position, so that it does not get stuck holding a loaf of bread.

Monitoring and diagnostics enable early detection of problems. For Mary's gripper, she ultimately implemented pressure sensors that could detect decreasing suction force before loaves would fall.

Concrete Improvements with Measurable Results

The concrete improvements that Marie implemented as a result of the FMEA analysis were:

Filtered air supply to prevent crumbs from entering the pneumatic system. This reduced the likelihood of blockages from 7 to 3.

Suction force sensors that continuously monitored the performance of the gripper and issued warnings when performance declined. This improved the detectability from 7 to 2.

Automatic cleaning system that blows out the air channels every 2 hours to prevent the accumulation of crumbs. This reduced the probability further to 2.

Redundant gripper as a backup system, reducing the severity of failure from 8 to 4, because production could continue.

The combined effect of these improvements reduced the RPN from 392 to 16 (2 × 4 × 2), an improvement of more than 95%. After these improvements, the packaging machine at Tromp Bakery ran for 18 months without major failures.

4.5 Practical FMEA Implementation

The implementation of FMEA in real design projects requires practical methodologies that fit within project constraints of time, budget, and availability of personnel.

Timing and Scope Definition

Timing is crucial for FMEA. Too early in the design process and there are still too few details known to perform meaningful analyses. Too late and the design decisions are already locked in, making improvements expensive and difficult. The right timing is when the concept design is finalized, but the detailed elaboration can still be adjusted.

Scope definition determines which components of the system are analyzed and to what level of detail. Mary uses the rule that FMEA is performed down to the level of "replaceable units" – components that are replaced as a whole during maintenance.

Scoring Guidelines for Consistency

Scoring guidelines ensure consistency in assessments across different team members and projects. Mary uses the AIAG standard:

Probability Scoring (Occurrence):

1: Very unlikely (less than 1 in 1,500,000)
2: Unlikely (1 in 150,000)
3: Very low (1 in 15,000)
4: Low (1 in 2,000)
5: Moderately low (1 in 400)
6: Moderate (1 in 80)
7: Moderately high (1 in 20)
8: High (1 in 8)
9: Very high (1 in 3)
10: Almost certain (more than 1 in 2)

Ernst scoring (Severity):

1: No effect (customer notices nothing)
2: Very minor effect (minor inconvenience)
3: Minor effect (customer slightly irritated)
4: Very minor effect on performance
5: Minor effect on performance
6: Moderate effect on performance
7: Significant effect on performance
8: High effect (system unusable)
9: Very high effect (safety risk with warning)
10: Catastrophic (safety risk without warning)

'Detectability Scoring (Detection):'

1: Almost certainly detected (automatic detection)
2: Very high probability of detection (multiple checks)
3: High detection probability (effective checks)
4: Moderately high detection probability (good controls)
5: Moderate detection probability (standard checks)
6: Low detection probability (limited checks)

7: Very low detection probability (irregular checks)
8: Small detection probability (weak inspections)
9: Very low detection probability (no systematic checks)
10: Not detectable (no inspections)

She realized that these descriptions still left plenty of room for interpretation, but decided to simply start with this list and clarify the definitions during the sessions as needed.

4.6 Common Mistakes in FMEA

Mary's experience with FMEA implementation across various projects had taught her to recognize and avoid certain common mistakes.

Starting too late is one of the most common mistakes. FMEA is often seen as a checkbox that needs to be checked for project approval, rather than a design tool. When FMEA is only performed after the design has already been largely finalized, improvements can no longer be implemented.

Insufficient team diversity leads to incomplete analyses. FMEA performed by designers alone misses the practical insights of users and maintenance personnel. Conversely, FMEA conducted solely by operational staff lacks the technical depth needed for proper root cause analysis.

Focus on technical failure modes only, while many failures are caused by human errors, inadequate procedures, or management decisions. Mary learned to include human factors in her analyses as well.

Perfectionism where teams attempt to analyze every conceivable failure mode, including those with negligible probability or impact. This leads to "analysis paralysis" and undermines the value of FMEA as a prioritization tool.

No follow-up on recommended actions where FMEA is performed to meet obligations, but the identified improvements are never implemented. Mary therefore implemented systematic tracking of all FMEA actions through to successful completion.

Static thinking where FMEA is seen as a one-time activity. In practice, systems, environments, and usage patterns change. FMEA must be updated when significant changes occur.

4.7 Summary: The Core Lessons

FMEA transforms how engineers think about reliability by providing a systematic, team-based approach to identifying and preventing potential failure problems. The Risk Priority Number (RPN) system combines probability, severity, and detectability to prioritize improvement actions. Industrial studies show that systematic FMEA implementation reduces design-related failures by 60-80%.

Effective FMEA requires diverse teams and practical implementation – it is not a bureaucratic exercise but a powerful design tool. A thorough FMEA for a critical subsystem requires 8-16 hours of team time, but prevents an average of 200-400 hours of problem-solving during operation.

For every reliability engineer, FMEA is an indispensable skill that makes the difference between systems that merely function and systems that function reliably. In the next chapter, we will learn how design for maintainability ensures that these optimized systems can also be maintained efficiently.

5 Design for Maintainability

5.1 The Story of Site Manager Pedro Vargas

Rick rubbed his temples and stared at the maintenance report on his desk. As a senior engineer at the chemical company ChemTech, he had seen this moment coming for years, but had hoped it would hold off a little longer. The main pump of their polymerization reactor needed to be replaced, but the design made this practically impossible without a major and costly operation.

"We told you people this five years ago," maintenance chief William muttered as he walked into Rick's office. "That reactor was designed by people who never held a wrench in their hands. First those valves that we could only get out by dismantling half the piping. Now this pump. You just can't get to it."

Pedro nodded wearily. He had been there himself when the engineering team had designed the reactor five years ago. Everyone had focused on technical performance and cost control. The warnings from the maintenance department about accessibility had been noted, but under time pressure they had been treated as "nice to have" rather than essential requirements.
"What's this going to cost?" Pedro asked, though he already dreaded the answer. Will opened his folder. "A minimum of two weeks of downtime. We first have to dismantle part of the piping, then rent a mobile crane to get the old pump out, and then rebuild everything. I estimate $180,000 in direct costs, plus $420,000 in production losses due to the downtime."

Pedro suddenly realized he had made a critical mistake five years ago. He had focused on building a technically advanced system, but had barely thought about how it would be maintained once it was up and running. Maintainability was treated as an afterthought rather than as a design criterion. This painful experience would become the beginning of a different approach to technical projects.

5.2 Maintainability as a Design Criterion

Maintainability is the degree to which a system can be effectively and efficiently maintained to keep it in good condition or restore it to a functional state. It is not only about being able to repair defects, but also about performing preventive maintenance, inspections, calibrations, and upgrades.

Good maintainability manifests itself in short maintenance times, simple procedures, minimal system downtime, and low costs for spare parts and tools. Poor maintainability leads to long repair times, complex procedures, extended downtime, and high maintenance costs.

The Critical Time Window

As described in chapter 1, the 80/20 rule applies here as well: maintainability is largely determined during design. For ChemTech's reactor, this meant that the decision to use the main pump placed directly against the wall — a decision made to save space during construction — would lead to many additional hours of maintenance time over the lifetime of the installation. The $3,000 that had been saved on floor space would no longer outweigh the maintenance costs after just one month.

The Business Case for Maintainability

Studies show that design decisions determine 60-80% of maintenance costs over the asset's lifetime. An investment of 1% extra in maintainability during design and construction — or simply conducting an additional analysis and improving the design without extra investment — can lead to 10-15% lower costs during operations over the lifetime of an asset.

For a chemical reactor with a service life of 25 years, this means that $42,000 extra spent on maintainability during construction
$2.1 million in operating costs. Often it's not even about expensive additions, but about smarter design choices: access panels in the right places, lifting points where you need them, or workspace that simply moves a fence one meter further away. These seemingly small choices can have the greatest impact, because they prevent days of downtime when repairs are

needed. One smart design choice can avoid days of downtime per year. Maintainability is not just a technical choice, but a strategic investment.

5.3 The Nine Elements of Downtime

When a system fails, the total downtime consists of several elements. Understanding these elements helps in taking targeted measures to minimize total downtime.

Administrative delay is the time between the occurrence of a failure and the moment a technician begins diagnostics. This includes the time to report the failure, create work orders, assign technicians, and set priorities.

Logistical delay is the time required to gather tools, spare parts, documents, and other necessities before the actual work can begin.

Access time is the time required to safely gain access to the defective part. This includes shutting down systems, obtaining work permits, and clearing access routes.

Diagnosis time is the time required to identify the exact cause of the problem. The better the built-in diagnostics and documentation, the shorter this time.

Time to obtain parts is the time to obtain the correct spare part when it is not in stock.

Active repair time is the actual time to replace or repair the defective component. Ironically, this is often the smallest part of the total downtime.

Test time is the time required to verify whether the repair was successful and the system is functioning normally again.

Startup time is the time required to restart the system and allow it to stabilize at normal operating parameters.

Administrative closure is the time to close out work orders, to update documentation and record lessons learned.

For a previous failure of ChemTech's main pump, analysis showed that the actual repair accounted for only 25% of the total downtime. The remaining 75% was overhead largely caused by poor maintainability. By improving accessibility and documentation, this time could be reduced to eight hours — cutting downtime costs in half.

This analysis reveals an important principle: most downtime is not caused by the actual repair work itself, but by the preparations and consequences surrounding it. This is precisely where good design for maintainability can make the greatest difference.

5.4 The Five Design Principles for Maintainability

Based on his experience at ChemTech and similar projects, Rick identified five design principles that can prevent 80 percent of maintainability problems. These principles are fundamental, because they determine whether a system actually maintainable in practice, regardless of how good the individual components are.

Accessibility

Accessibility forms the foundation of every maintainable system. All components that require maintenance must be easily reachable without disassembling other equipment. Without accessibility, even the simplest maintenance tasks become complex and time-consuming.

Practical Implementation at ChemTech:

- Minimum 80 cm of working space around equipment that requires regular maintenance
- At least 120 cm for equipment that requires complete disassembly
- Components that are inspected daily at a maximum height of 180 cm
- Heavy components equipped with lifting points with access for lifting equipment

These guidelines resulted in 40 percent shorter maintenance times and significantly fewer work-related incidents. Technicians could work much more efficiently because they were no longer hampered by hard-to-reach components or lack of workspace.

Visibility

While accessibility lays the physical foundation, visibility ensures that technicians can work effectively. Technicians must be able to see what they are doing and assess the condition of components without disassembly. This prevents unnecessary disassembly and reduces the likelihood of errors during maintenance.

Practical Implementation at ChemTech:

- Transparent windows in enclosures for visual inspection
- LED lighting at all maintenance points, switched with local switches
- Color-coded piping according to DIN 2403 standard
- Clear component labels with tag numbers

The result was a 25 percent reduction in diagnostic time and fewer components being unnecessarily disassembled for inspection. Technicians could identify problems faster and had greater confidence in their diagnoses due to the improved visibility.

Standardization

Standardization of components, tools, and procedures significantly reduces the complexity of maintenance. This third principle builds on the first two by ensuring that accessible and visible components can also be maintained in a uniform way.

Practical Implementation at ChemTech:

- All bolts M12 or M16 where mechanically possible
- Standard DIN flanges instead of special designs
- All electrical connections with standard industrial connectors
- Grease fittings of the same type throughout the entire installation

The inventory value of spare parts decreased by 60 percent through this systematic approach. Lead times improved due to higher turnover rates, and there was a dramatic reduction from 27 different fitting sizes to 4 standard sizes — a reduction of 85 percent. This had not only financial benefits, but also resulted in fewer errors when ordering and installing parts.

		Pressure class		
		PN6	PN10	PN16
Size	DN10	X	X	X
	DN15	X	X	X
	DN20	X	X	X
	DN25	X	X	X
	DN32	X	X	X
	DN40	X	X	X
	DN50	X	X	X
	DN65	X	X	X
	DN80	X	X	X

		Pressure class		
		PN6	PN10	PN16
Size	DN10			
	DN15			X
	DN20			
	DN25			X
	DN32			
	DN40			
	DN50			X
	DN65			
	DN80			X

Figure 5-1 Number of fitting sizes without (left) and with (right) standardization

Advantages and Challenges of Standardization

Standardization offers considerable advantages in the area of spare parts efficiency, because fewer variants mean higher turnover rates and lower inventory costs. In addition, concentrating purchasing volume with fewer suppliers yields concrete benefits such as volume discounts, better service, and greater willingness from suppliers to parts to keep in stock. A third important advantage is the use of proven technology, which inherently has lower failure risks than new developments.

At the same time, standardization also brings challenges. Engineers may be forced to make suboptimal technical choices in order to stay within the established standards. Strict standardization can also delay the implementation of improved technologies, because new solutions must first be evaluated and approved within the standardization framework. Finally, worldwide standards can be locally suboptimal due to differences in availability of parts, service organizations, or environmental factors. The art is finding the right balance between uniformity and flexibility.

Modularity

Modularity builds on the first three principles by designing systems in interchangeable modules that can be maintained independently. Instead of complex on-site repairs, modules can defective modules can be quickly replaced and repaired elsewhere under optimal conditions.

Practical implementation at ChemTech:

- Standard baseplates with pre-drilled mounting points
- Quick couplings for process piping that can be disconnected and reconnected without tools
- Electrical plug-in connectors for motor connections
- Interchangeable subassemblies for critical components

This resulted in 60% shorter downtime during maintenance and better repair quality. Because modules could be repaired under controlled conditions in the workshop rather than in the often cramped and noisy production environment, the quality of the repair work also improved significantly.

Built-in diagnostics

The fifth principle integrates intelligence into the system. Modern sensor technology can monitor the condition of components and provide early warnings before problems become critical, thereby planned maintenance becomes possible instead of reactive repair.

Practical Implementation at ChemTech:

- Vibration sensors on all rotating equipment
- Temperature sensors on bearings and motors
- Pressure sensors in hydraulic systems
- Vibration monitoring for pumps to detect cavitation
- Runtime monitoring on all moving components

Rick also significantly improved the failure messages that the system generated. The failure messages indicated in both plain language and technical codes exactly what was happening. The operator screen visually showed the precise location of the problem, which further reduced diagnosis time and prevented unnecessary searches for the cause of failures.

These five principles work synergistically together: accessible and visible components can be effectively standardized, standardized components make modular designs practically feasible, and built-in diagnostics maximize the effectiveness of all preceding principles by enabling predictive maintenance.

5.5 Documentation and Knowledge Transfer

Even the best-designed system is difficult to maintain without good documentation. This documentation must cover several levels of detail of coverage: from quick reference cards for daily tasks to detailed guides for troubleshooting complex problems.

Drawings That Match Reality

As-built drawings are essential, but they are often not updated during construction. At ChemTech, Rick made sure that all changes during construction were systematically documented.
The system included:

- Digital marking tools that allowed assembly technicians to indicate changes directly
- Daily rounds during which all changes were checked
- Version Control for Consistency Between Documents
- QR codes on equipment for direct access to drawings via tablets

Maintenance Procedures That Work

Maintenance procedures must provide step-by-step instructions for all routine tasks. These procedures must be tested by the people who will actually be using them.
For ChemTech, Pedro Vargas developed procedures according to a standardized format:

- Scope and safety (what is being done and what precautions are taken)
- Tools and materials (complete list of requirements)

- Step-by-step instructions with visual aids
- Quality checks (verification of correct work)
- Cleanup Procedures

Troubleshooting Guides with Decision Trees

Troubleshooting guides help with diagnosing problems. The best guides use a decision tree approach: "If symptom X occurs, check Y first, then Z." An effective troubleshooting guide uses:

- Clear symptom descriptions
- Logical sequence of checks
- References to relevant procedures
- Escalation paths when standard solutions don't work

Training and Competency Development

If the people performing the maintenance don't know how to handle the system, they can't maintain it either. Training is an important part of maintainability, not an optional extra.

Operator Training for First-Line Maintenance

Operator training is important, because operators are usually the first to notice problems. They need to know what normal behavior looks like, how to recognize problems, and what initial actions they can take.
For ChemTech, Rick developed an operator training program with:

- System Fundamentals (How the Process Works)
- Normal operations (normal sounds, vibrations, temperatures)
- Abnormal Conditions (How to Recognize Problems)
- Routine Maintenance (Simple Tasks for Operators)
- Routine maintenance (simple tasks for operators)

Training for Technicians in Advanced Maintenance

Training for technicians must include both theoretical background and practical skills. Technicians must understand how the system works, not just how to replace parts.

The training program included:

- Process Chemistry (Basic Understanding of Processes)
- Equipment Operation (How Machines Interact)
- Diagnostic Techniques (Systematic Problem Identification)
- Repair Procedures (Step-by-Step Instructions)
- Safety Protocols (Work Permits, Lockout/Tagout)

5.6 Safety and Regulations

Maintainability must always take safety aspects into account. Systems that are dangerous to maintain are often poorly maintained, because technicians avoid risks or take shortcuts.

Lockout/Tagout Procedures

Lockout/tagout procedures must be simple and reliable. Complex procedures are often not followed, which can lead to accidents. For ChemTech, Rick developed a locking system with:

- Centralized lockout points for each subsystem
- Visual verification that systems are actually isolated
- Standardized locks and labels
- Clear procedures that are easy to follow

Integrated Work Permits

Work permits for hazardous work must be integrated into the maintenance process, not treated as a separate bureaucratic layer. The system must be logical and add real safety value.

Environmental factors such as temperature, noise, vibrations, and chemical exposure must be taken into account in the design of locations where people need to perform maintenance. For ChemTech, this meant installing sound insulation around compressors and adequate ventilation at all maintenance points.

5.7 Common Mistakes and Pitfalls

From Jack's experience, several common mistakes emerged that other project managers can avoid:

Too tight a schedule for maintainability during project execution is a common occurrence. Maintainability is seen as "nice to have" rather than "necessary," which means that under time pressure, compromises are made on accessibility and documentation.

Insufficient involvement of maintenance personnel during design and implementation. The people who will actually maintain the system have valuable insights that are often not utilized, because they are not involved early enough.

Focus on acquisition costs instead of life cycle costs.
Decisions that save a few thousand dollars during construction can cost tens of thousands of dollars extra during operation.

Underestimating the documentation effort. Good as-built documentation and procedures require considerable time and money, but this investment pays back many times over in more efficient maintenance.

Insufficient training budget. Training is often viewed as a cost rather than an investment, while inadequate training leads to significantly higher operational costs due to errors and inefficiency.

5.8 Summary: The Core Lessons

Maintainability is not something you can add to a system after the fact — it must be incorporated from the very beginning as a design criterion. The nine elements of downtime demonstrate that active repair time is often only a small part of the total downtime.

By deliberately working on the five design principles — accessibility, visibility, standardization, modularity, and built-in diagnostics — you can significantly reduce the total maintenance burden. The business case is compelling: investments inmaintainability during design and construction pay for themselves many times back in lower costs during operations.

According to industrial studies, organizations that systematically focus on maintainability show an average improvement of 20-40% in maintenance performance. This improvement translates into lower costs, higher availability, and improved safety.

For Rick and ChemTech, it was too late to fully prevent the current problems, but the experience taught them the value of systematic attention to maintainability. Their next project would be designed with maintenance in mind from the very beginning, with the result that the system not only performed better, but also cost far less to keep running.

The story of Rick illustrates that investing in maintainability is not only technically sound, but also economically necessary. In a world where operational peak performance is becoming increasingly important for competitive advantage, good maintainability is no longer a luxury but a requirement.

In the next chapter, we will learn how RCM (Reliability Centered Maintenance) in projects helps us develop systematic maintenance strategies that build on the maintainability we have built into the design.

6 RCM in Projects

The Story of Project Manager Karen

Karen stared at the schedule for her latest project: a completely new production line for pharmaceutical tablets. As an experienced project manager, she had delivered dozens of installations, but this project was different. Management had explicitly asked for "reliability from day one" and wanted to prevent the maintenance budget from spiraling out of control in the first years.

Karen knew why this was important. In earlier projects, the first period after commissioning had always been problematic: equipment that ran flawlessly during trial runs failed unexpectedly during production; maintenance intervals proved too optimistic and the team was repeatedly caught off guard by failures. The previous project — a packaging line — had already generated $180,000 in unplanned maintenance costs in the first six months. For a pharmaceutical environment with strict quality requirements, that was unacceptable.

"This time it has to be different," Karen decided. Instead of organizing maintenance after the fact, she wanted to integrate it from the design phase onward. She knew Reliability Centered Maintenance (RCM) primarily from aviation and its application to existing installations, but she wondered: why not apply that methodology directly during the design phase?

6.1 Why RCM in Projects?

RCM originated in the 1960s in aviation, where traditional preventive maintenance programs became unsustainable. The essence is simple: determine the consequences of failure for each component and develop the most effective maintenance strategy.
Many companies apply RCM only to existing installations, even though it was originally developed for analysis during the design phase. Karen saw the opportunity to apply this during the design phase, so that

maintenance concepts could be built in from the very beginning and freedom of choice in design and suppliers could be fully utilized.

As discussed in Chapter 4, FMEA identifies failure modes, causes, and effects. RCM builds on this and adds a crucial step: developing maintenance strategies for failure modes that cannot be fully prevented. Where FMEA primarily tries to avoid design flaws, RCM also focuses on operational reality and cost-benefit tradeoffs.

For Karen, this meant that she not only knew what could fail, but also how to best prevent or manage it.

6.2 RCM Methodology in Brief

RCM works according to seven systematic core questions that lead to the optimal maintenance strategy:

1. What are the functions of the system?
2. How can the system fail?
3. What are the causes of failure?
4. What are the effects of failure?
5. What are the consequences of failure?
6. What preventive measures are possible?
7. What if no preventive measure is possible?

RCM uses a structured decision diagram that leads you step by step to the optimal maintenance strategy. The logic is clear and always follows the same systematic approach.

The first question the decision diagram asks is: 'Are the consequences of this failure hidden?' Hidden failures only have consequences when a second failure occurs (such as safety systems that fail). Evident failures have immediately noticeable consequences.

For evident failures, RCM then asks the question: are the consequences acceptable according to your risk matrix? This creates three main paths:

1. Hidden Failures
2. Evident and Acceptable
3. Evident and Unacceptable

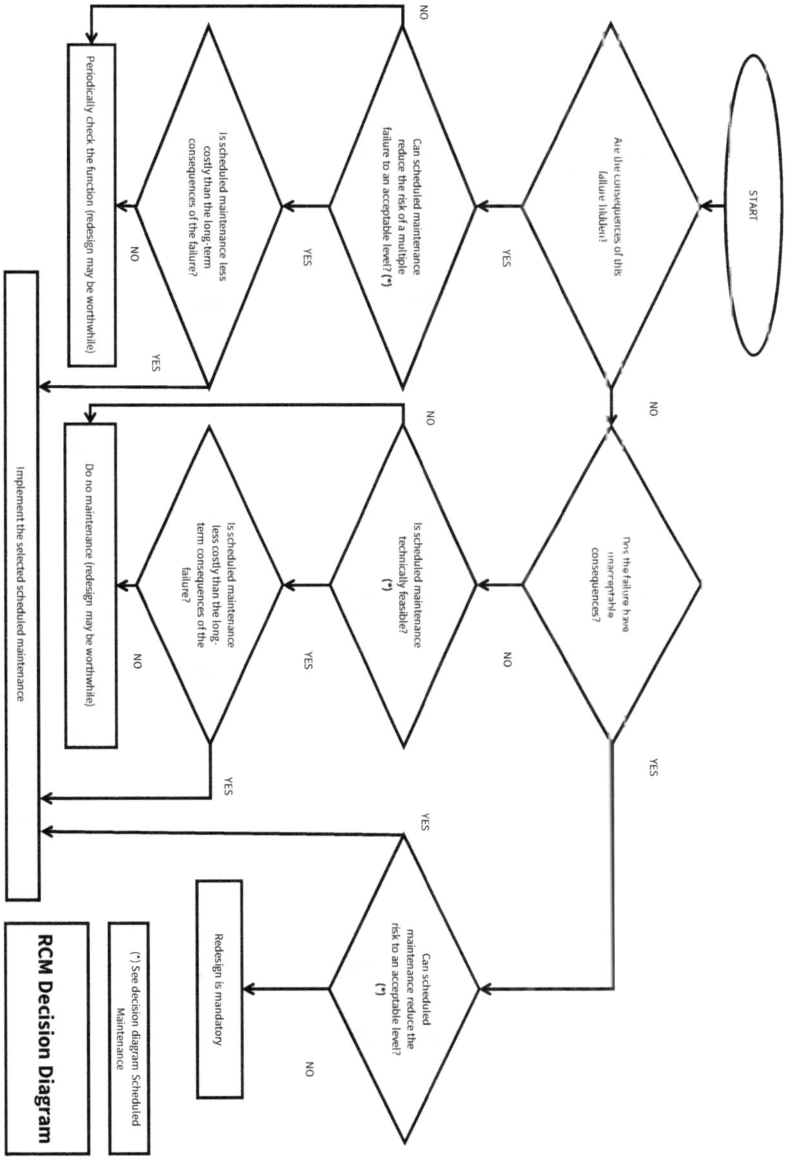

Figure 6-1 Main decision diagram of RCM

For each path, RCM goes through the same two questions:

Is condition-based maintenance possible? Can you determine through measurement or inspection determine that the failure is developing? RCM favors this approach because you perform maintenance at exactly the right time, have lower costs, and gain more insight into actual failure behavior.

Is time-based maintenance possible? Periodic replacement based on a fixed interval. Only if condition monitoring is not feasible.

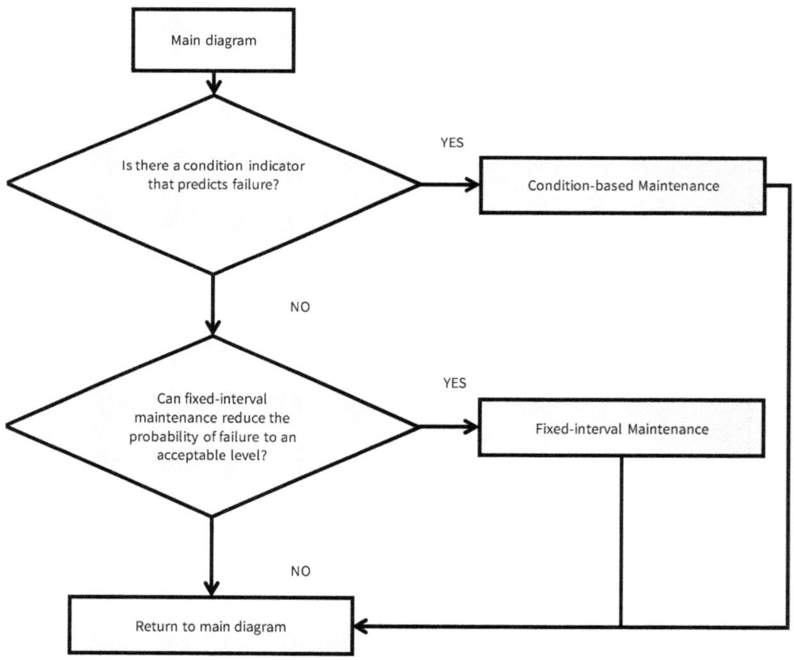

Figure 6-2 Sub-decision diagram of RCM

Then the conclusion differs slightly depending on the path.

Evident and unacceptable: If no maintenance is possible, then redesign is mandatory. Safety and critical aspects take priority over cost. If maintenance is possible, perform the chosen maintenance.
Evident and acceptable: If no maintenance is possible, do nothing. You then consciously choose to take action only after a failure has already occurred. This is called 'failure-based maintenance'. It is then worth considering modifying the design so that the failure cannot occur, or occurs less frequently. If maintenance is possible, but costs more per year than the failures that occur when no maintenance is performed, the conclusion is also that it is best not to perform preventive maintenance: 'run to failure'.
Hidden failures: If maintenance is not possible or is too expensive, you can perform function tests. You then check, for example, whether your emergency stops are still working. The difference with condition-based maintenance is that with the latter you try to observe in advance whether a failure is going to occur.

RCM prevents both over- and under-maintenance by always putting safety first, choosing the most cost-effective strategy, considering realistic alternatives, and systematically justifying every decision.

6.3 Selective Application of RCM

RCM is time-consuming and expensive. That is why Karen only applies RCM in two specific situations. First, for assets that are both critical and where you have little experience. A new, advanced extruder warrants RCM because failure costs $3,500 per hour and because you don't know what maintenance is optimal. A steam boiler, on the other hand, does not get RCM, even though it is critical. After decades of experience, we know exactly what maintenance is required.

Second, for assets of which you have a large number. If you have 50 identical pumps, a thorough RCM analysis is worthwhile even if you save only $50 per pump per year. The total savings of $2,500 per year then justify the investment in analysis.

Criticality Ranking: How Do You Determine Which Assets Are Critical?

Before Karen could decide which assets deserved RCM, she first needed to know which assets were critical in the first place. Because applying RCM to non-critical assets is a waste of time and money. That is why she carried out a criticality ranking, which you read about in Chapter 3.

She assembled a multidisciplinary team of six people: a maintenance engineer who understood each machine technically, a process engineer who knew how failures rippled through the process, a production supervisor who could assess which machines were truly indispensable, and an experienced operator with the best feel for which equipment was problematic. A facilitator led the sessions to keep discussions focused.

In three sessions of two to three hours each, they systematically went through all 147 assets. For each asset, they asked the same question: what happens if this fails? They assessed the impact on:

- Patient safety – risk of contamination or dosing errors
- Production criticality – line shutdown
- Quality – Impact on Product Specifications
- Economic impact – high repair costs or extended downtime

An extruder causing $3,500 per hour in downtime costs received a high score. A spare pump with minimal impact received a low score.

The team did not fill in how often failures occur – after all, it depends on the maintenance plan that still needs to be developed. They assessed only: if it fails, what are the consequences?

After three sessions, the results were clear: 23 of the 147 assets turned out to be critical for one or more aspects. These 23 were eligible for further analysis according to the criteria described above.
In practice, only 5% of all assets receive full RCM. The remaining 95% receive Industrial RCM (a faster variant) or an individual approach by the reliability engineer.
This selectivity is crucial: RCM on all assets is cost-prohibitive and unnecessary. Focus on where it truly makes a difference.

6.4 Practical Application in Projects

Karen started the RCM analysis after the first detailed design, but before the construction of the installation. This gave her enough freedom to implement design changes without significant cost increases.
Karen focused on the 23 critical components. A major advantage of RCM in the project phase is that outcomes can directly lead to design improvements:
Tablet coating system: Instead of frequent manual cleaning, an automatic flushing system was installed. This cost an extra $25,000, but saved $180,000 over five years.
Conveyor belts: By choosing a modular system instead of one long belt, the replacement time was reduced from 12 hours to 30 minutes.
minutes. Extra design costs of $45,000 were already recouped at the first major failure.
Main motor tablet press: The main motor proved so critical that multiple strategies were combined: vibration monitoring, periodic lubrication, annual inspections, and thermal monitoring. The investment of $15,000 yielded $60,000 in annual savings.
Unlike traditional projects, Karen now had a complete maintenance plan ready at handover, built up in three levels:

- Daily care by operators: visual inspections, measurements and functional tests, integrated into the work

- Maintenance by the in-house maintenance department: weekly and monthly tasks such as lubricating, calibrating, and inspecting
- Specialized interventions: annual overhauls and replacements, with fixed intervals for critical components

6.5 Results, Lessons, and Recommendations

Eighteen months after commissioning, the results were convincing. The number of shutdowns was less than half of what they had expected, maintenance costs were 40% lower than anticipated, and product quality was consistently high. These successes yielded valuable insights for future projects.

The most important lesson was to integrate RCM early in the project cycle to retain the greatest design freedom, ideally after the detailed design but before construction, so that modifications are still possible without major cost increases. Multidisciplinary teams proved essential for better decision-making because different perspectives lead to more comprehensive analyses. Karen discovered that pragmatism is more important than perfectionism: 80% of the value comes from 30% of the analysis, and teams must focus on failure modes that truly matter to avoid "analysis paralysis," where endless analysis takes place without moving to action.

Design modifications typically delivered the greatest gains because they resolve structural problems rather than treating symptoms. At the same time, change management proved to be critical because even the best maintenance plans fail without buy-in. That is why it is important to involve operators early in the process and to clearly communicate what benefits RCM delivers for their day-to-day work.

6.6 Summary

RCM in the project phase delivers reliability from day one, lower maintenance costs and better team engagement. Where FMEA primarily analyzes, RCM adds maintenance strategy and design impact.

Karen's experience shows that a systematic, pragmatic approach leads to sustainable results. By selectively applying RCM to critical and unique assets, or to assets of which you have many, you get maximum return on the investment in analysis time.

Ideally, this is applied during the design phase, when you can still influence design choices that can prevent costly maintenance.

7 Life Cycle Costing

The Story of Procurement Officer Sandra

Buyer Sandra at a mid-sized machine shop faced a difficult decision. She had to choose between two pumps for the new cooling water system. Pump A cost $15,000 and had an efficiency of 78%. Pump B cost $22,000 but had an efficiency of 85%.
The purchasing director made it clear that budgets were tight and asked why she didn't just go with the cheaper option.

Sandra decided to run the numbers. The pump would operate 6,000 hours per year and consume 15 kW. At a rate of $0.12 per kWh, the 7% efficiency difference represented a savings of $1,140 per year. Over a lifespan of 15 years, pump B would therefore save $17,100 in energy costs – more than enough to offset the price difference over the entire lifetime. Her company, however, required a payback period of five years for investments. Based on that, she would have to choose the cheaper pump. But she had not yet looked at maintenance costs, availability and other factors.

"There must be a systematic way to make these kinds of decisions," Sandra thought. She was right. That method is called Life Cycle Costing, and it can make the difference between investments that create value for years and investments that turn out to be money pits.
But first Sandra had to solve a fundamental problem: how do you obtain reliable cost data for your Life Cycle Costing analysis? Now that you have mastered FMEA, you can generate reliable cost data by systematically analyzing what can go wrong and what it costs. And with the maintainability principles from chapter 5, you can realistically estimate maintenance costs.

7.1 The Problem of Short-Sighted Decision-Making

The tendency to focus on purchase price is understandable but costly. Capital budgets are visible and closely monitored by management. Operational costs disappear into various cost centers and become less directly linked to original decisions. The result is a systematic underestimation of life cycle costs.

Stacking of Safety Margins

A common problem Sandra encountered was what she called "the stacking of safety margins." A chemist specified that a dosing pump had to deliver 30 m^3/hour. The process engineer calculated with 33 m^3/hr "to be on the safe side." The mechanical engineer designed for 38 m^3/hr. The pump manufacturer supplied a standard model of 40 m^3/hr. In practice, the pump was throttled to 25 m^3/hr, because otherwise the process would be disrupted.

This oversizing had three negative consequences. First, the pump was more expensive than necessary. Second, it consumed more energy, because it was operating outside its optimal operating point. Third, it had a shorter service life due to problems caused by throttling.

Lack of Systematics

Sandra discovered that many organizations do have maintenance data, but it is spread across different systems. Energy costs are not recorded per asset. Downtime costs are not systematically recorded at all. Without this data, it is difficult to make well-informed Life Cycle Costing analyses.

Industrial research shows that organizations that implement systematic LCC achieve on average 15-25% lower life cycle costs than organizations that focus only on purchase price. For Sandra's company, with an annual investment budget of $2 million, this meant a potential savings of $300,000 to $500,000 per year.

Organizational separation

The separation between procurement and engineering on the one hand, and production and maintenance on the other, reinforces shortsighted decision-making. Buyers are evaluated on cost savings at the time of purchase. Engineers are assessed on technical performance and project lead time. Production and maintenance managers have no influence over procurement decisions, yet must live with the consequences.

Sandra experienced this firsthand. Her performance indicators were based on the percentage of cost savings relative to the original budgets. If she were to choose the more expensive pump, it would negatively affect her evaluation, even though it was the best choice for the company as a whole.

7.2 LCC Methodology: Cradle-to-Grave Thinking

Life Cycle Costing maps all costs from the initial concept through the final decommissioning costs. This holistic approach enables understanding the true impact of design choices and objective comparison of different alternatives.

Design costs

Design costs form the starting point of the life cycle. These include not only engineering hours, but also costs for feasibility studies, prototype construction, testing, and verification. For complex systems, design costs can represent 5% to 15% of total project costs. Good engineering in this phase prevents costs that are many times higher later.

For Sandra's pump system, the design costs were relatively limited, since it was a standard application. But she realized that a thorough analysis of the actual requirements could have prevented her from now facing this dilemma.

Construction Costs

Construction costs are the most visible element and include material, labor, transport, and commissioning. A common mistake is underbudgeting commissioning and training. A poor startup can increase life cycle costs significantly through infant failures that are never fully resolved.

Sandra calculated that the installation costs for both pumps would be approximately equal. The price difference of $7,000 was purely in the pump itself, not in the surrounding infrastructure.

Operational Costs: The Largest Cost Item

Operational costs typically represent the largest portion of LCC and include five main categories.

Personnel costs for operators and maintenance technicians typically rise annually with inflation plus productivity improvement. For Sandra's pumps, personnel costs were limited, since the system would operate largely automatically.

Energy costs are often the dominant factor and vary significantly between design options. For example, the life cycle costs of an electric motor consist of 95% to 99% of energy costs. For Sandra's pumps, this was the biggest difference. Pump A would cost $13,846 in energy annually. Pump B would cost $12,706. The difference of $1,140 per year was substantial, and over 15 years this became $17,100 — and that without accounting for energy price increases.

Maintenance costs often differ considerably between alternatives. Through the FMEA analysis from chapter 4, Sandra was able to predict that Pump A, which would be throttled to avoid disrupting the process, was expected to cost $800 per year in maintenance. Pump B, which would operate closer to its optimal operating point, would cost approximately $600 per year.

Downtime costs are difficult to predict, but can exceed all other costs. Sandra's reliability analysis showed that pump A, due to its higher load and worse operating point, would fail more often — approximately every three years versus every four years for pump B. Each failure would cost approximately $3,000 in production loss and repair.

Decommissioning Costs

Decommissioning costs are often overlooked, but can be substantial, especially for systems containing hazardous materials. For Sandra's cooling water system these costs were limited, but she still took them
$1,000 per pump in her calculation for responsible disposal of materials.

7.3 Practical Tools for LCC Analysis

The creation of reliable LCC analyses requires a systematic approach and the right tools. Sandra developed a methodology that you can apply directly in your own projects.

Net Present Value Calculations

NCW calculations are essential, because money in the future is worth less than the same amount today. The formula is

$$NCW = -Investment + \Sigma(Cashflow/(1+r)^t)$$

where r is the discount rate (required return) and t is the number of years. A cost stream of $1,000 over 15 years, such as disposal costs, at an 8% discount rate has a present value of only $315. Sandra used a discount rate of 8%, based on her company's cost of capital. For the energy costs, this meant that the annual savings of $120 over 15 years had a present value of $1,028.

The choice of discount rate has a significant impact on outcomes. Sandra conducted a sensitivity analysis with interest rates of 6%, 8%, and 10%. At 6%, the difference between the pumps became larger; at 10%, smaller. This helped her understand how robust her conclusion was.

Sensitivity Analysis for Robust Decisions

Sensitivity analysis tests how robust your conclusions are against changes in assumptions. Sandra varied key parameters such as energy prices, maintenance costs, and service life by ±20% and examined how this affected the outcome.

The analysis showed that energy prices had the greatest impact on her decision. If energy prices were to rise 20% higher than expected, the advantage of pump B would become significantly greater. If they were to remain 20% lower, the difference would shrink, but pump B would still be attractive.

Break-even analysis

Break-even analysis identifies the point at which two alternatives

have equal costs. For Sandra's pumps, this meant: when would the additional purchase costs of $7,000 be offset by lower operating costs?
With the annual savings of $1,310 ($1,140 energy + $50 maintenance + $120 reduced failure costs), the break-even point would be reached after 5.3 years. Since the expected service life was 15 years, this was comfortably within the lifespan.

7.4 Input for Calculations and Data Challenges

One of the greatest challenges in LCC analyses is obtaining
reliable input data. Without good data, your calculations are only as good as your guesses. Sandra developed a systematic approach for collecting data.

Data Quality Warning

Now that you have mastered FMEA, you can generate reliable cost data by systematically analyzing what can go wrong and what it costs. With the maintainability principles from chapter 5, you can realistically estimate maintenance costs. These methodologies give you the tools to generate better input data than traditional estimates.

Enterprise Asset Management Systems

EAM systems often contain valuable information, but it is not always directly usable. Failure records are often incomplete or inaccurate. Maintenance costs are not always correctly allocated to assets. Downtime data is often completely missing.
Sandra discovered that her company did have data on energy consumption, but it was only measured at the building level, not per machine. For maintenance costs, she did have data per asset, but it had not been recorded consistently. She first had to invest three months in cleaning up the data before she could make reliable analyses.

Expert Interviews for Missing Data

Expert interviews can compensate for data shortages. Experienced technicians and operators often have an intuitively sound understanding of asset behavior. "This pump fails about three times a year" and "Every failure costs us an average of half a day of production" are valuable inputs for analyses.

Sandra interviewed four experienced technicians about their experiences with different pump types. Their estimates for failure frequencies aligned well with the limited data she had. For new technologies where no experience existed yet, their assessments based on comparable systems were very valuable.

Keep in mind that people tend to weigh recent events more heavily than historical averages. Sandra corrected for this by interviewing multiple people and comparing their estimates with historical data where available.

Supplier Data and Benchmark Studies

Supplier data can help with new technologies where own experience is lacking. Ask not only about technical specifications, but also about experience data from other customers. How much maintenance does the machine require on average? What are the typical failure patterns? Which spare parts have the shortest service life?

Sandra asked both pump suppliers for reference visits. At pump A she saw an installation that had been running for five years and had indeed achieved the predicted maintenance costs. At pump B she visited a newer installation that had only been running for two years, but where the owner was very satisfied with the performance.

7.5 Sandra's complete LCC equation

With all the data together, Sandra could complete her full analysis. She combined the FMEA insights on failure modes with the maintainability principles to create realistic cost estimates.

Acquisition Costs and Installation

Pump A cost $15,000, Pump B cost $22,000. The difference of $7,000 was purely in the pump itself. Installation costs were for both pumps equal, because they used the same connections and foundations.

Energy costs over 15 years (discounted at 8%)

Pump A would cost $13,846 per year in energy. Discounted over 15 years at 8%: $118,609.
Pump B would cost $12,706 annually in energy. Discounted over 15 years: $108,846. Pump B therefore saved $9,763.

Maintenance Costs Based on Reliability Analysis

Through the FMEA analysis, Sandra was able to predict that both pumps would have comparable maintenance costs due to their similar operating points.
Pump A: $7,704 discounted ($900/year).
Pump B: $7,276 discounted ($850/year).
Here too, pump B saved costs: $428.

Failure Costs from Reliability Data

Pump A was expected to fail every 2.5 years (6 failures over 15 years). Pump B every 3.5 years (4.3 failures). She estimated the cost per failure at: $3,000.
Pump A failure costs discounted: $11,088.
Pump B failure costs discounted: $7,948.
Savings pump B: $3,140.

Total LCC Equation

Pump A total:
$ 15,000 + $ 118,609 + $ 7,704 + $ 11,088 + $ 315 = $ 152,716
Pump B total:
$ 22,000 + $ 108,846 + $ 7,276 + $ 7,948 + $ 315 = $ 146,385

Conclusion and Sensitivity Analysis

Pump B saved $6,331 over its lifetime, despite the $7,000 higher purchase price. The lower operating costs more than compensated for the extra investment, making pump B the economically better choice.
Sandra decided to perform a sensitivity analysis to see what would happen to her conclusion if a number of things developed differently than she had assumed. Her analysis showed that her conclusion was robust against

changes in key parameters. Even under unfavorable scenarios, pump B remained the more cost-effective option:
At energy prices that would remain 15% lower than expected, pump B would still retain an advantage of $2,800. Only at energy prices that would remain 20% lower — an unlikely scenario — would pump B's advantage disappear.
With a shorter service life of 12 years instead of 15 years, pump B
$4,100 cheaper overall. Even with a 10-year lifespan, pump B would still save $2,600.

If both pumps were to unexpectedly have identical maintenance costs, pump B would still be $6,200 more cost-effective due to energy savings alone.
The analysis confirmed that pump B would remain the better choice under virtually all realistic scenarios. Only extreme assumptions—such as energy prices permanently dropping by 20% or a service life shorter than 8 years—could reverse the preference. Given the trend of rising energy prices and the expected service life of industrial pumps, pump B was not only economically superior but also the more future-proof choice.
This robust analysis gave Sandra the confidence to recommend,' Sandra said, 'knowing that the decision held up under various realistic scenarios.

7.6 Remaining Service Life Analysis and Replacement Strategies

Many assets in industry and infrastructure are older than their design life. Organizations therefore struggle with fundamental questions: should we replace assets, or can they keep running for a while longer? When is the right time for replacement? Choosing the optimal replacement timing is crucial, because wrong decisions lead to unnecessarily high costs or unacceptable risks.

The Dilemma of Replacement Timing

If you replace assets too early, you pay unnecessarily high interest and depreciation costs. In many organizations, the costs of interest and depreciation on investments are higher than the operational costs. In most countries, tens of thousands of kilometers of pipelines for gas, water, and sewage are in the ground. Network companies can save enormous amounts of money if they keep these pipelines in service for a few more years before replacing them. However, if they replace them too late, maintenance costs and risks increase significantly.

The problem is complicated by the fact that replacement decisions are often not driven by technical considerations alone. Sometimes the requirements that companies or regulators place on assets change. In other cases, technological developments or innovations are reason to invest in new assets, even if the current assets are still functioning well. Newer assets may operate faster, at lower cost, or in a more environmentally friendly manner.

Systematic Approach to Remaining Service Life Analysis

The first question in a remaining service life analysis is whether the installation has critical failure modes. Can something go wrong that is unacceptable according to the business values? Which components can fail due to aging with serious consequences?

Using a criticality ranking, which you read about in Chapter 3, you determine which assets are critical. These are the assets whose failure has unacceptable consequences. In practice, only a small portion of the assets in a system are truly critical. With a Failure Mode and Effect Analysis, you then determine which failure modes can occur in these critical assets. This FMEA provides insight into the precise causes that can lead to failure.

If these causes are time-dependent, such as corrosion or fatigue, it is necessary to periodically assess the condition of the assets. This technical examination of the condition reveals whether the assets are still suitable for continued use. Because the failure modes are time-dependent, it is usually also possible to indicate what the approximate remaining service life is.

Data Analysis for Targeted Assessment

To more purposefully determine which assets need to be investigated, an increasing number of companies are conducting data analyses. Gas pipelines can be made of various materials: steel with welded joints, gray cast iron, nodular cast iron, asbestos cement, rigid PVC, impact-resistant PVC, and polyethylene. There are pipelines in all kinds of diameters, laid in all kinds of soil types.

With the help of data analysis, you gain better insight into, for example, the percentage of steel pipes that have corroded after 20 years in sandy soil versus clay soil. You can also map out the maintenance costs by age, material, and diameter. If you know what the maintenance costs are for a pipe that is 40 years old, you can estimate what the maintenance costs will be in 10 years for pipes that are currently 30 years old.

If causes are not time-dependent and inspection reveals no failure-predicting parameter, then it is not possible to determine a remaining service life. The probability of failure is then still just as high as when the asset was new. In that case, periodic inspection is needed to continue monitoring the condition.

CAPEX versus OPEX Development Over Time

Even systems that are not critical have a finite lifespan, of course. As they age, it can happen that they begin to show more failures. If the failures are not critical, this is by definition acceptable. It is not a problem if a small hole develops in a pipe, as long as you can weld it shut again and the consequences are not significant. However, if holes gradually begin to appear more and more frequently, it may be more economically sensible to replace the entire pipe all at once.

To properly weigh this, you need to set CAPEX and OPEX against each other. The CAPEX of an installation appears in the accounting records as interest and depreciation. These costs are highest in the first years after investment and decrease as the asset gets older and is further depreciated.

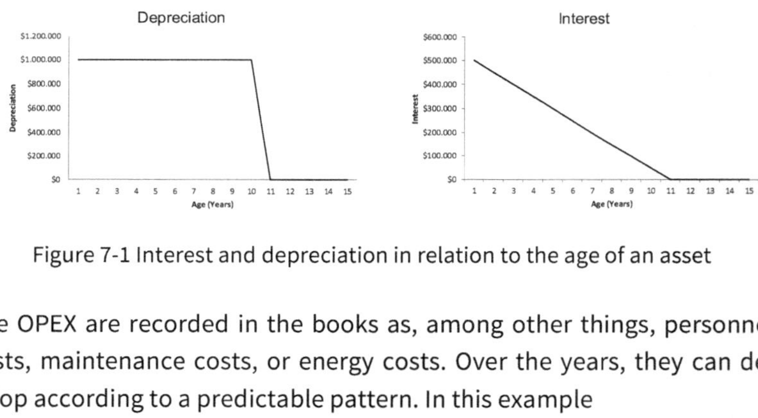

Figure 7-1 Interest and depreciation in relation to the age of an asset

The OPEX are recorded in the books as, among other things, personnel costs, maintenance costs, or energy costs. Over the years, they can develop according to a predictable pattern. In this example maintenance costs increase after the 10th year by 2% per year and energy costs from the 15th year onward by 1.5% per year due to reduced efficiency of aging equipment.

These numbers are merely examples. The actual numbers vary greatly depending on the quality and condition of the components.

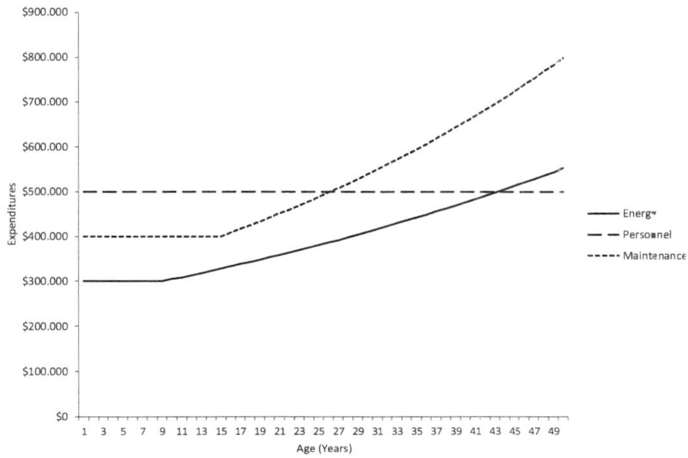

Figure 7-2 Operational costs increase over the years

Determining the Optimal Replacement Moment

To determine the optimal service life, you need to add up all costs adding up and averaging over the lifetime. You thus also average the costs of interest and depreciation over the entire lifetime. After all, if assets remain in service longer, the organization needs to invest less. For the calculation, you therefore depreciate the acquisition of the assets over their actual lifetime, not the depreciation period that companies typically use in their accounting.

At a certain point, however, maintenance costs can rise so high that it is cheaper to buy a new machine. In a real-world example, a machine cost an average of $2,000,000 per year when the company replaced it every 15 years. When the company operated it for 35 years, these costs dropped to $1,700,000. If the company continued even longer, however, costs rose again due to exponentially increasing maintenance costs and increased failure risks, see figure 7.3.

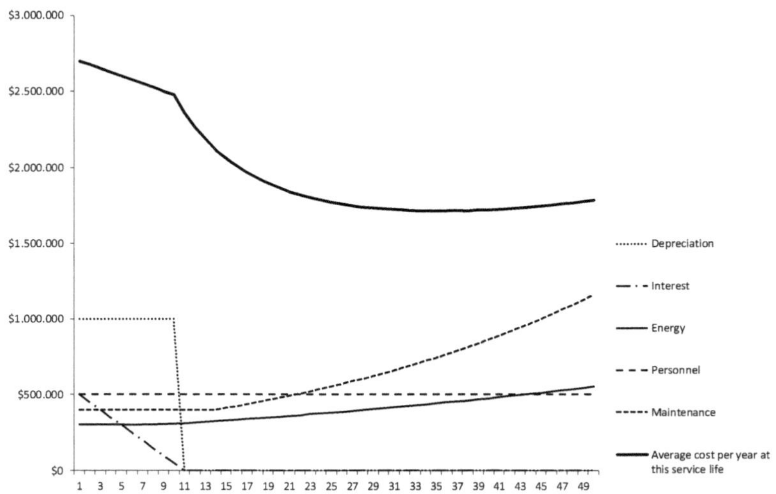

Figure 7-3 Optimal service life depends on many factors

The optimal replacement frequency is where the average annual costs are lowest. This is the point where the additional costs of using the asset for one more year equal the savings from postponing the new investment by one more year.

Practical Implementation of Replacement Strategies

It is usually not possible to perform these calculations in detail at the time an installation is being built. However, they can be performed after the assets have been in use for a number of years. The data from the EAM is then extremely valuable in analyzing maintenance costs. This data often needs to be cleaned up before an analysis is possible, because many organizations do not properly track their failures and the replacement timing of components.

For some applications, you can also rely on interviews with people who are well acquainted with the installation. Based on their experiences, you can generally gain a good understanding of the behavior of the installation and make realistic estimates of future developments.

Of course, it is important to carefully consider whether it is truly necessary to replace the entire installation in order to bring maintenance costs down. Often, a few small measures are sufficient to reverse the majority of the cost increase. A thorough technical analysis can demonstrate which components genuinely need to be replaced and which ones can last for many more years.

Integration with Other LCC Considerations

Remaining service life analysis must be integrated with the other LCC methodologies discussed in this chapter. The NPV calculations from section 7.3 help discount future replacement costs to their present value. The sensitivity analysis from section 7.4 shows how robust replacement decisions are against changes in assumptions about future cost developments.

Business case development from chapter 8 becomes crucial when replacement decisions need to be made. Management must be convinced that the timing of replacement has been optimized and that both premature and delayed replacement prove financially disadvantageous.
The criticality analysis from chapter 3 helps prioritize replacement decisions. Critical assets deserve more frequent evaluation and possibly more proactive replacement than non-critical assets where a run-to-failure strategy may be acceptable.

Risk Management in Replacement Strategies

Replacement decisions must take into account risks that go beyond purely economic optimization. Reputational risks, safety risks, environmental risks, and strategic risks can all argue for earlier replacement than would be purely economically optimal.
A systematic risk assessment helps factor in these considerations. You can apply risk analysis methodologies to quantify the impact of potential failure scenarios and incorporate them into the replacement decision.
By systematically applying remaining service life analysis, organizations transform their asset portfolio management from reactive

troubleshooting to proactive optimization. The result is lower total life cycle costs, better risk management, and more predictable operations.

7.7 Digital Support for LCC

Modern digital twin technology makes it possible to optimize life cycle costs through virtual modeling of different scenarios. This technology supports LCC methodologies by enabling better prediction of operational costs, but does not replace the systematic analysis of cost categories.
IoT sensors can monitor real-time energy consumption and maintenance needs to support condition-based maintenance strategies. AI and machine learning support cost prediction through pattern recognition in historical data, but do not replace the expertise of engineers in performing LCC analyses.

7.8 Summary: The Power of Systematic Thinking

Sandra's story illustrates why Life Cycle Costing is so powerful. Without systematic analysis, she would likely have chosen the cheaper pump and later regretted the higher operational costs. With LCC, she was able to make a well-informed decision that took all relevant factors into account. The key lessons are clear.
A short-sighted focus on purchase price systematically leads to suboptimal decisions that cost organizations an average of 15% to 25% more over the asset's lifetime.

Second, LCC methodology offers a structured way to include all costs over the lifecycle. The formula

$$\text{NCW} = \text{Investment} + \Sigma(\text{Cashflow}/(1+r)^t)$$

helps with correctly discounting future costs.
Third, sensitivity analysis helps make robust decisions despite uncertainty. By varying parameters by ±20%, you can see how stable your conclusions are.

Fourth, good data is crucial but often well worth the effort to obtain. Through FMEA (chapter 4) and maintainability analysis (chapter 5), you can generate more reliable cost data than traditional estimates.

Fifth, industrial studies make clear that organizations that systematically implement LCC consistently realize lower life cycle costs and make better investment decisions.

Sandra's pump analysis illustrates the power of Life Cycle Costing. By systematically analyzing all costs over the entire lifecycle, you prevent suboptimal decisions based on purchase price alone.
The LCC methodology has now shown you how to compare different options and calculate the true costs. These analyses are essential for what comes next: building a compelling business case. In the following chapter, you will learn how to combine LCC data, FMEA insights, and risk analyses into a financial justification that convinces management to invest in reliability.

8 Building the Business Case

The Story of Controller Michael

Michael stared at the cost calculation on his screen and shook his head. As financial controller at a large food manufacturing plant, he had just received a proposal from the maintenance team for an investment of $205,000 in vibration sensors and predictive maintenance software.
"This absolutely cannot happen," he mumbled to his colleague. "We have no budget for this kind of experimentation."

The proposal had been put together technically by chief mechanic Henry. The sensors would be able to predict when pumps and motors needed maintenance, which would reduce unplanned downtime. The software could analyze trends and determine optimal maintenance intervals. It all sounded very modern and forward-thinking.
But for Michael, the essential element was missing: a convincing financial justification. The proposal did contain some general statements about "reduced downtime" and "lower maintenance costs," but no concrete figures, no risk analysis, and no comparison with alternatives.
"How am I supposed to explain this to management?" Michael asked himself. "They'll want to know about the payback period, the return on investment, and the certainty of the benefits."

The frustration was mutual. Henry didn't understand why such a clearly useful improvement had to be made financially complicated. To him, the logic was straightforward: better monitoring means fewer failures, fewer failures means lower costs. Why did that have to be wrapped up in complex spreadsheets with NPV calculations and risk analyses?
This conflict between technical necessity and financial justification plays out daily in many companies. Excellent reliability projects fail because the people requesting them cannot demonstrate that they create value for the organization. At the same time, investments receive approval that appear financially attractive but are technically risky.

The solution lies in developing solid business cases that clearly articulate both the technical necessity and the financial value.
Michael and Henry had to work together to tell a story that that was both technically sound and financially convincing.

8.1 ROI Calculation of Reliability Investments

After weeks of discussion between Michael and Henry, a breakthrough finally came when they decided to systematically map out the financial impact of reliability problems together. What they discovered surprised even Michael: the true costs of unreliability were much higher than either of them had suspected.

Systematic Cost Inventory

The first step was creating a complete inventory of all costs related to asset failures. Michael and Henry discovered that traditional accounting records captured only a fraction of the actual costs.
Direct repair costs were visible in the records: parts, labor from in-house technicians, and external services. For the main pumps in their plant, these costs averaged $85,000 per year. But this was merely the tip of the iceberg.
The costs of production loss were much greater but less visible. Every hour of downtime cost $12,000 in lost margin, but these costs were not systematically recorded per failure. By combining maintenance reports with production data, Michael was able to calculate that pump failures caused 47 hours of downtime annually, amounting to $564,000 in production losses.
Quality costs arose because after every failure, the production process had to be recalibrated. The first two hours after restart, they produced substandard product that had to be rejected or reworked. Michael calculated that this amounted annually to $95,000 was lost in material losses and additional processing time.
Overtime and emergency measures cost $23,000 per year in premium rates for technicians who had to resolve failures outside of regular business hours, plus the costs of rush parts deliveries and hiring specialists.

ROI Calculation Methodology

With this complete cost picture, Michael was able to make a reliable ROI calculation for Henry's proposal. The methodology he developed later became standard for all reliability investments in their organization.

Calculation of Cost Overview

Investment costs:	
'Sensors and software:'	$ 180,000
Installation and training:	$ 25,000
Total investment:	$ 205,000

Current annual failure costs:	
Direct repair costs:	$ 85,000
Production loss (47 hours × $12,000):	$ 564,000
Quality costs:	$ 95,000
Overtime and emergency measures:	$ 23,000
Total per year:	$ 767,000

For the annual savings, Michael used a conservative approach based on experience data from comparable projects. Predictive maintenance was expected to prevent 60% of unplanned failures. Of the 47 hours of annual downtime, they could therefore prevent approximately 28 hours.

Michael calculated the annual savings as follows:

Less production loss: 28 hours × $12,000 =	$ 336,000
Reduced repair costs	$ 35,000
Reduced quality costs:	$ 40,000
Reduced overtime and emergency measures:	$ 15,000
The total annual savings amounted to	$ 426,000

With a discount rate of 8% over a period of 7 years (the expected lifespan of the system), Michael calculated a net present value of $2,034,000. The return on investment came out at 208% per year. The payback period at less than half a year. This ROI may seem unrealistically high, but it is what we see time and again in reliability investments. For a relatively small investment, companies can get much more out of their existing assets. That is why the ROIs are much higher than you may be used to from investments in new assets.

8.2 Accounting for Risks in Calculations

Michael realized that this calculation was too optimistic, because it did not account for uncertainty and risks. He therefore developed a model that incorporated various scenarios and risks.

The optimistic scenario (25% probability) assumed that the system would perform even better than expected, with a 75% reduction in unplanned failures. The realistic scenario (50% probability) held to the 60% reduction. The pessimistic scenario (25% probability) assumed only 35% reduction due to implementation problems and operator resistance.

The weighted average annual savings came to $368,000, which yields a return of 180% per year. This approach, which accounted for the risks, gave management greater confidence in the robustness of the business case.

Sensitivity Analysis and Scenario Planning

To further strengthen the business case, Michael conducted a comprehensive sensitivity analysis that showed how changes in key assumptions would affect the outcomes.

The analysis showed that downtime costs had the greatest impact on the ROI. A 20% underestimation of downtime costs would reduce the return to 460%, still excellent. A 20% overestimation would increase the return to 700%.

The effectiveness of the system was the second most important factor. Even if the system could only prevent 40% of failures instead of the expected 60%, the return would still amount to 350%.

8.3 A Standard Model for Presenting the Business Case

Michael developed a standardized presentation format that was later used for all reliability investments. The format consisted of five main sections that told a logical story.

The problem statement described the current situation, the costs of failures, and the risks of inaction. For the pump project, Michael showed that the current costs were $767,000 per year and were rising due to aging equipment.

The proposed solution described the technical proposal, the implementation strategy, and the expected results. Michael used simple, non-technical language to explain the benefits to management that had no engineering background.

The financial analysis presented all calculations in a clear, visual format. Charts showed cash flows over time, the payback period, and cumulative savings. Tables provided detailed breakdowns of costs and benefits.

The risk analysis discussed the key uncertainties and how they were managed. Michael presented not only the risks but also the measures that had been planned to mitigate them.

The recommendation concluded with a clear call to action, including the requested approval, implementation timeline, and next steps.

Quick Methods for Risk Assessment

The development of fast but reliable methods for risk assessment proved essential for making compelling business cases. Michael discovered that management was skeptical about business cases that did not account for risks and uncertainties.
Michael used the criticality ranking and risk matrix you read about in Chapter 3 to clarify the consequences of failure of the main pumps. This also allowed him to convince management when an investment yielded less financial return, but which did reduce the risks for aspects such as safety, environment, quality, or sustainability to an acceptable level.

8.4 Summary: The Power of Compelling Business Cases

Setting up a solid business case transforms technically excellent reliability projects from "nice ideas" into "must-have investments" through systematic financial justification.
Michael's experience shows that the combination of thorough cost analysis, industry benchmarks, and projections that account for risks gives management the confidence to invest in reliability.

The most important lessons are clear. First, every business case must begin with a complete inventory of all failure costs, not only the visible repair costs. Production loss, quality problems, and indirect costs are often more than ten times greater than direct costs.
Second, an analysis that accounts for risk is essential for robust business cases. By incorporating optimistic, realistic, and pessimistic scenarios, you can demonstrate that investments remain valuable even when results fall short of expectations.

Third, a standardized presentation format makes business cases consistent and comparable. This helps management make informed decisions when setting priorities among different investment opportunities.

With a well-supported business case, you get more projects approved than with arguments alone. For Michael's organization, this meant $2.3 million in additional reliability investments over three years.

9 Commissioning and Operational Readiness

You have created a reliable design (chapters 4–6), the business case has been approved (chapter 8), and the installation has been ordered and built. Now the truth comes to the surface: does it really work? Commissioning is where theory becomes practice. It is the phase in which the design is tested, operators are trained, and the organization prepared for operation. Technically perfect installations can be disastrous fail if commissioning is not performed properly. This chapter shows you how to systematically achieve operational readiness.

The Story of Manager Frank

Frank stood in the control room of the new biomass power plant and looked at the impressive row of screens displaying the entire process live. After two years of construction, the installation was finally ready for use. All systems had been installed, all Factory Acceptance Tests had been successfully completed, and the Site Acceptance Tests had demonstrated that every component was working according to specifications.

Yet Frank felt a knot in his stomach. As an experienced manager, he had witnessed too many times how technically perfect installations could fail disastrously during the first months of operation. The problem rarely lay in the technology itself, but in what he had come to call "the human factor": were the operators truly ready to operate this complex installation? Were the maintenance technicians trained for the specific challenges of biomass combustion? Was the organization prepared for the transition from construction project to operating plant?

The direct cause of Frank's concerns was an incident from three days earlier. During a test, one of the biomass conveyor belts had jammed — a relatively simple problem that could have been resolved in a few minutes. But it took two hours to fix the failure, simply because nobody knew exactly where the reset button was located or which procedure to follow. "We built a technical masterpiece," Frank said to his team, "but forgot to create an operating organization." This realization would become the foundation for a fundamentally different approach to commissioning, where operational readiness became just as important as technical performance.

9.1 Commissioning: The Three Aspects Simultaneously

Traditional commissioning focuses on verifying that all systems are working technically correctly: pumps run in the right direction, sensors give the correct readings, and safety systems respond as designed. But Frank realized that this was only part of the story. Real commissioning is about converting a construction project into a running plant.

This conversion has three aspects that must be addressed simultaneously:

- **Technical aspect:** all traditional activities (pre-commissioning checks, loop testing, performance testing, and system tests)
- **Business aspect:** building skills (training of personnel, creating procedures, learning work processes)
- **Organizational aspect:** transition from project to operations (maintenance organization, reporting lines, division of responsibilities)

For Frank's biomass power plant, commissioning took on a much broader scope than planned. The starting point was a thorough inventory that went beyond technical lists and also included soft factors: teamwork, communication, and decision-making. Frank discovered that all three aspects must occur simultaneously. Teams learn to work together while systems are being tested and procedures are being tried out.

9.2 Systematic Planning

Effective commissioning requires methodical planning that begins in the design phase of a project for all three aspects. Frank learned this when his first plans proved too optimistic, because they were based on the assumption that everything would work as designed.

The revised approach began with developing a philosophy that defined the goals, principles, and approval criteria. For the biomass plant, the goal

was not merely "getting all systems to work," but "delivering a plant that can operate reliably, safely, and cost-effectively from day one."
This philosophy was translated into a detailed plan that described all activities, responsibilities, and milestones. The plan was divided into five phases, each with specific entry criteria, activities, and exit criteria.

The Five Phases of Commissioning

Preparation encompassed all preparatory activities: they checked the quality of the installation, calibrated instruments, and functionally tested individual components. This phase was important, because problems they missed here would be far more expensive and time-consuming to resolve later.
Cold commissioning tested all systems without process material: the team filled and tested water circuits, energized electrical systems, and tested control loops. In this phase they discovered integration issues and coupling problems, before the complexity of actual process operation was added.
Hot commissioning introduced process material and tested all systems under real operating conditions. For the biomass plant, this meant they actually burned biomass and tested all processes under full load.
Performance testing verified that all performance guarantees were met and that the system could operate sustainably. This included not only technical metrics such as efficiency and emission levels, but also business metrics such as availability and safety performance.

Operational handover was the final phase in which the team transferred full responsibility to the production department. This phase included final training sessions, documentation transfer, and the establishment of ongoing support relationships.

9.3 Assessment of Operational Readiness

One of the most important tools Frank developed was a methodical operational readiness assessment that evaluated all aspects. This assessment went far beyond traditional technical punch lists and also encompassed people, processes, and organizational aspects.
The assessment was structured around eight core areas.

1. Technical readiness evaluated whether all systems were installed, tested, and operational. This involved not only checking whether components functioned according to specification (quality), but also whether the installation quality met the class requirements from chapter 2.7. This included verification of critical aspects such as shaft alignment, operating point optimization, piping integrity, and all other factors that determine the service life of components. On rotating equipment, they performed standard vibration measurements.
2. Personnel readiness assessed whether sufficient trained personnel were available for all operational roles.
3. Procedural readiness verified that all required operating procedures had been developed, tested, and implemented.
4. Maintenance readiness examined the readiness of the maintenance organization: were spare parts available, had maintenance plans been developed, and had technicians been trained?
5. Supply readiness evaluated whether all required materials and services were available.
6. Management system readiness assessed whether all management processes were functioning.
7. Safety readiness was a vital area that evaluated whether all safety systems were functioning and whether personnel were adequately trained in safety procedures.
8. Business readiness, finally, looked at the commercial aspects: were all contracts in place, were financial systems operational, and had all permits been obtained?

For each of these areas, Frank developed detailed lists with specific criteria. Technical readiness was not assessed as a simple yes/no question, but as a percentage of completeness in which both functionality and installation quality (class) were systematically validated against the requirements from the specification phase.

9.4 Training and Skills Development

One of the most important lessons Frank learned was that technical training alone was not enough. Operators needed to not only know which buttons to press, but also understand why, and what to do when things didn't go according to plan.

For the biomass power plant, Frank developed a comprehensive training program that encompassed three levels. Basic training gave all employees a fundamental understanding of the processes, safety procedures, and their role in the organization. Role-specific training focused on the specific skills required for each role: operators learned process control, maintenance technicians learned specific repair procedures, and managers learned performance monitoring. Advanced training prepared important employees for abnormal situations and emergency procedures.

But training was more than just knowledge transfer. Frank realized that team building and developing effective communication were just as important. The team integrated the training program into the commissioning schedule, so that personnel could gain hands-on experience Frank implemented a mentorship program in which experienced operators and technicians from other plants took temporary assignments to support new teams during startup.

9.5 Procedure Development and Documentation

Frank discovered that procedure development was an important part of commissioning that is often underestimated in traditional projects.

For the biomass power plant, they developed four types of procedures. Normal operating procedures described all routine activities: startup, shutdown, normal operation, and routine maintenance.

Abnormal operating procedures covered all foreseeable deviations: alarms, minor failures, and non-conforming conditions.

Emergency procedures described the response to serious incidents: fire, explosion hazards, personal injury, and environmental discharges. Maintenance procedures covered all planned maintenance tasks: inspections, preventive maintenance, and scheduled replacements.

The team developed each procedure according to a standardized format that provided step-by-step instructions, included safety warnings, and defined clear acceptance criteria. They tested procedures during commissioning and adjusted them based on practical experience. Procedure validation formed an important component in which not only experts reviewed procedures, but the actual users performed them under realistic conditions.

9.6 Integrated Testing and Reliability Validation

While individual components had been tested during FAT, the team had to verify the integration of all systems during commissioning. For complex installations such as the biomass power plant, this often represented the most challenging phase of commissioning.

Frank developed a structured approach to integrated system testing that began with simple loops and gradually built up to complete process tests. They tested control loops first individually, then integrated them into subsystems, and finally tested them as a complete process. The team tested safety systems extensively under all relevant conditions. Interface testing between different systems was particularly important — they tested every interface methodically to ensure that information exchanged correctly.

Factory and Site Acceptance Tests

Suppliers conduct FATs in their controlled environments, and these are ideal for extensive reliability testing. For the biomass plant, Frank developed a FAT program that went far beyond traditional functional testing. The protocol included extensive burn-in testing in which all systems ran under normal load for hours to eliminate early-life failures before transport to the site.

SAT, on the other hand, tests the integration of the system in the actual operating environment. Frank's SAT protocol focused on aspects that could only be tested in the real environment: interface testing with existing systems and verification of performance under actual operating conditions. Frank developed a methodical approach in which they performed key reliability tests during FAT to identify problems before systems were shipped to the site.

Reliability Testing During Commissioning

Commissioning offers a unique opportunity for reliability testing that is not possible during FAT, but is important for long-term performance. One of the most valuable tests is methodical burn-in testing to eliminate early-life failures before the system enters commercial production. Commissioning also offers the opportunity to test systems under the actual environmental stress they will experience during operation.
Frank conducted failure mode tests in which they simulated various component failures to verify that the system would exhibit gradual degradation and that operators would receive adequate warnings. These tests frequently identified critical single points of failure that they could address by adding redundancy or implementing backup procedures before commercial operations began.

9.7 Reliability Growth and Lessons Learned

Frank discovered that commissioning was not just a testing phase, but also a reliability improvement phase. During commissioning, the team finds and resolves many "teething problems," which significantly improves the long-term reliability of the installation.
For the biomass power plant, Frank implemented a structured program during commissioning. They carefully documented and analyzed all failures and near-misses. The team identified root causes and implemented corrective actions. They measured reliability growth by tracking key KPIs during commissioning: number of failures/week, mean time between failures, and system availability.

Frank introduced a formal problem reporting system in which they documented all issues, regardless of severity, and tracked them through to resolution. They implemented a methodical lessons-learned process in which they documented and analyzed all commissioning experiences. The team used these lessons to improve future activities and to identify design improvements for future projects. They documented and analyzed all reliability tests methodically and used test results to develop maintenance procedures, refine operator training, and validate emergency procedures.

9.8 Handover and Performance Monitoring

The transition from commissioning to normal operations is an important moment that requires careful planning. Frank developed a structured handover process that ensures a smooth transition without loss of continuity.

The handover process began with a comprehensive system walkthrough during which they methodically inspected and documented all systems. The team identified and prioritized all outstanding items. Documentation transfer formed a key component, during which they formally transferred all technical documentation, operating procedures, maintenance procedures, and training materials to the operating organization.
They held knowledge transfer sessions where commissioning engineers shared their experiences with operations personnel. Frank introduced a gradual transition approach in which commissioning engineers remained available for consultation during initial operations. They gradually reduced this support as operations teams gained confidence and experience.

Frank realized that successful commissioning did not end at handover, but that methodical performance monitoring and support of employees during the first months after startup were necessary. They implemented a comprehensive performance monitoring system that tracked all key performance indicators. The team introduced continuous improvement processes in which they methodically analyzed operational experiences and fed them back to design teams for future projects. They established feedback loops to incorporate lessons learned from commissioning into future projects.

9.9 Summary: What Really Works

Commissioning is not about testing machines alone. It is about preparing people, procedures, and organizations. If you only test technical aspects, you get problems like at Frank's biomass plant: a jammed conveyor belt that sat idle for two hours, because nobody knew how to fix it.

The most important lessons are simple. Start planning early — not only when everything has been built. Don't just check whether the technology works, but also whether people know what they need to do. Train your staff well, not only technically but also in collaboration. Create clear procedures and actually test them. Make sure everyone knows who is responsible for what.

Frank's approach works: good preparation for commissioning leads to a faster startup, fewer problems, and better performance. It does take time and money, but you'll recoup that investment many times over.

The essence: commissioning succeeds not only when you build perfect machines, but also when you create perfect teams that can operate them. That is the difference between a technical marvel that fails and a plant that runs well from day one.

In the next chapter, we will learn how supplier selection and procurement can be optimized for reliability, building further on the business case methodologies to make total cost of ownership decisions.

10 Supplier Selection and Procurement

The Story of Buyer Anna

Anna looked at the three quotes on her desk. As a senior buyer at a mid-sized chemical plant, she had just received quotes for a critical pump system. At first glance, the choice seemed simple: supplier A offered the lowest price of $395,000, supplier B asked $445,000, and supplier C wanted $485,000.

But lead technician Bert had warned her that the lowest price wasn't always the best choice. "Five years ago we also went with the cheapest option," he told her. "Those pumps have cost us more in maintenance and downtime since then than we saved on the purchase. On top of that, we got almost no support from the supplier when problems arose."
The problem was that Anna's procurement procedures were based on traditional criteria: price, lead time, and technical specifications. This approach worked well for standard purchases such as office supplies or raw materials, but failed with complex technical systems where the total lifecycle costs were far more important than the purchase price.

"How do I determine which supplier truly offers the best value?" Anna wondered. "And how can I ensure that we don't just get a good pump system, but also stable support throughout its entire service life?"
This challenge would become the catalyst for developing a systematic approach to supplier selection that went beyond traditional procurement criteria and focused on the total value proposition for assets that must perform for years to come.

10.1 Procurement versus Project Management

Project management focuses on the selection of suppliers from a project perspective. Anna approaches it differently — she focuses on the procurement and contractual aspects of reliability. These two perspectives complement each other by addressing different phases of supplier management.

Where project managers look at the success of projects and timely delivery, Anna looks at years of performance and support. These different perspectives are both necessary to obtain assets that work well from day one and continue to perform for years to come.

Criteria for Supplier Evaluation

Anna discovered that traditional procurement criteria represented only a small part of the total value proposition. She needed to develop a comprehensive set of evaluation criteria for making sound decisions about suppliers.

Assessment of Technical Capabilities

The evaluation of technical capabilities went far beyond simply checking whether a supplier could meet the basic specifications. Anna developed a systematic approach that encompassed both current and capacities and future possibilities.

Expertise in design she evaluated by looking at the supplier's engineering team, their experience with similar applications, and their ability to deliver customization when needed. For the pump system, Anna asked for reference projects in similar chemical applications and the specific challenges that had been overcome in those projects.

Production capacities she assessed through plant visits, certifications for quality and production processes. Anna wanted to see that suppliers had modern manufacturing equipment, trained personnel and robust quality control systems.

Capacity for innovation she measured through investments in research and development, new product introductions, and partnerships with research institutions. For long-term assets it was important that suppliers invested in improving their technologies.

Track Record of Operational Reliability

Assessing the history of supplier performance required access to objective performance data, which was often difficult to obtain. Anna developed several methods to acquire this information.

Reference visits proved to be the most valuable source of information. Through direct conversations with maintenance managers and operators, Anna was able to assess the actual performance of equipment in the field. She asked specifically about failure frequencies, typical maintenance issues, and satisfaction with vendor support.

Data on warranty claims from suppliers themselves provided insight into the reliability of their products. Suppliers who were transparent about their warranty claim rates and willing to share this data demonstrated confidence in their products.

Databases from industry sources such as those from insurers, maintenance contractors, and trade associations sometimes contain operational reliability statistics for various suppliers and product lines.

Capabilities for Support and Service

The quality of support and service was often underestimated during supplier selection, but proved crucial to the total cost of ownership of complex assets.

Onsite presence was important for a fast response to problems. Anna evaluated whether suppliers had local service organizations, how quickly they could be on-site, and whether they had technicians available in-country for communication without language barriers.

Technical support she evaluated by the availability of application engineers, their response times for technical questions, and the quality of their documentation and training materials. Vendors with technical support hotlines that were available 24/7 and remote diagnostic capabilities scored higher.

Availability of spare parts and lead times were critical for minimizing downtime. Anna asked about stock levels of critical parts, average lead times, and guarantees of availability throughout the expected service life of the equipment.

10.2 Requirements for Dependability Contracts

Translating operational reliability requirements into concrete contract conditions proved to be one of the most challenging aspects of procurement. Anna learned that vague language in contracts led to disappointments and disputes, while overly specific requirements could deter suppliers.

Performance Guarantees and Penalties

Developing fair yet effective performance guarantees required a delicate balance between protecting the customer and making them acceptable to suppliers.

Availability guarantees specified the minimum percentage of time that equipment had to be operational. For the pump system, Anna set a guarantee of 98% availability during the first two years, excluding planned maintenance shutdowns. This was ambitious, but based on industry benchmarks for comparable systems.

Continuity guarantees defined maximum failure frequencies. Anna specified a maximum of 2 significant failures per year during the first 3 years, where significant failures were events that caused more than 4 hours of downtime.

Performance guarantees extended beyond operational reliability alone and included efficiency, capacity, and quality parameters. The pump system was required to achieve a guaranteed 95% of design efficiency and maintain this for a minimum of 5 years.

Mechanisms for sanctions had to be fair but effective. Anna developed a tiered approach in which minor breaches of warranties led to demerits for service, but significant performance issues could lead to cost reimbursements or even termination of contracts.

Provisions and Warranty Coverage

Comprehensive warranty provisions protected against unexpected costs during the early phase of equipment life.

Basic warranty coverage included all material and manufacturing defects for a minimum of 24 months. For critical components such as main bearings and seals, Anna required 36 months of warranty, because these components often did not show problems until after a longer period of operation.

Extended warranty options also covered wear parts and consumables for a specified period. Although more expensive, these options offered predictable costs and transfer of risk to the supplier.

Service for on-site warranty eliminated costs for transportation and downtime for repairs under warranty. For the pump system, Anna required that all warranty work be performed on-site by qualified technicians within 48 hours of notification.

Maintenance Support Requirements

Long-term support was crucial for complex assets that would remain in service for 20–30 years.

Guarantees for the availability of spare parts ensured that critical parts would remain available throughout the expected lifespan of the equipment. Anna required a 15-year availability guarantee for all parts that were not obtainable from third parties.

Service level agreements for technical support defined response times for different types of support requests. For critical issues, she required a four-hour response time for telephone support and 24 hours for on-site assistance.

Support for software became increasingly important as equipment became more computer-controlled. Contracts had to specify how long software updates would be available, at what cost, and what would happen if the vendor discontinued support for software platforms.

Template for Supplier Scorecard

Anna developed a scorecard that brought structure to the complex comparisons of suppliers. By systematically weighing different criteria, she could make objective decisions that went beyond simply comparing prices. With this scorecard, she awarded points to the suppliers for the following aspects:

- Technical
- Operational Reliability
- Service
- Financial
- Strategic

You can find this scorecard in Appendix A.

Scores between 8-10 indicate excellent suppliers who often deserve preference, even if they are not the cheapest.

Scores between 6-8 are acceptable, but require extra attention to weaker areas.
Scores below 6 are risky for critical applications.
Pay particular attention to large differences between categories. A supplier with high scores in technical areas but low scores for service can cause problems with support. Conversely, excellent service can compensate for weaker technical aspects in less critical applications.
Adapt this scorecard to your situation by adjusting the weightings and criteria to match the specific requirements of your organization and the type of equipment being purchased.

10.3 Training and Development of Competencies

Anna discovered that procurement focused on operational certainty required specific knowledge and skills that went beyond traditional purchasing training.

Training of Purchasing Staff

Buyers had to learn how to specify requirements for operational reliability, evaluate suppliers on aspects of continuity, and negotiate warranty terms that support stability. This required specific knowledge of reliability engineering principles.
Technical interfaces with engineering teams were crucial. Buyers needed to be able to communicate with engineers about technical requirements and understand the business impact of different technical choices.

Negotiating Contracts for Continuity

Negotiating contracts for reliability required a different approach than traditional procurement contracts. Anna learned that it was important to build partnerships rather than merely to conduct transactions.
Guarantees for performance had to be negotiated in a way that motivated suppliers to excel, not merely to meet minimum standards. This required creative incentive structures and shared-risk models.

10.4 Stability of finances and business continuity

As Anna gained more experience, she realized that assessing the financial health of suppliers was becoming increasingly important. As assets became more complex and required longer support, the risks of supplier problems increased.

Analysis of Supplier Finances

Analyses of supplier finances consisted of reviewing annual reports, credit ratings, and cash flow trends. Suppliers experiencing financial difficulties posed a risk to future support, even if their current products were excellent.
Market position and competitive advantage determined the likelihood that a supplier would remain successful in the future. Suppliers with strong market positions and a clear competitive edge were more likely to invest in product development and customer support.

Strategies for Risk Mitigation

Diversification of the supplier base reduced the risk of dependence on a single supplier. Anna developed strategies to keep critical components available from multiple sources.
Escrow Arrangements for Critical Software and Documentation
protected against supplier bankruptcy. These arrangements ensured that essential information remained available even if suppliers discontinued support.

10.5 Summary: The Core Lessons

Supplier selection and procurement for operational certainty requires a fundamentally different approach than traditional procurement. It is not just about obtaining the lowest price, but about creating partnerships that deliver value throughout the full lifecycle of assets.
The most important lessons are clear.

First, supplier evaluation must go beyond price and technical specifications to include performance history, service capabilities, and financial stability.

Second, requirements for operational certainty contracts demand specific performance guarantees, warranty provisions, and long-term support agreements that go beyond standard commercial terms.

Third, the evaluation must be systematic, with scorecard methodologies that enable objective comparison between suppliers across all relevant criteria.

Fourth, procurement focused on operational certainty requires specific competencies for procurement teams to be effective.
To apply these lessons, use the scorecard from Appendix A, which brings together all the criteria from this chapter.
Anna's experience shows that the investment in systematic
supplier selection processes pays off through better asset performance, lower total cost of ownership, and stronger supplier partnerships. By using the templates and methodologies from this chapter, organizations can transform their procurement from a cost center into a creator of value.

11 Asset Portfolio Management

Karen and the Power Plant Expansion

It was eight o'clock in the morning when project manager Karen stood at the power plant of Metal Precision Group with head technician Pete. The report he had with him would herald an important strategic decision for the company.

"The main boilers have reached the end of their service life," Pete began. "The corrosion is accelerating and the efficiency is declining. We have six months to make a definitive decision about what we want to do with them."

Karen looked at the boilers that had faithfully served the machine shop. "What are our options?"

"We have three options: continue operating with increased risk, renovate for $180,000, or complete replacement for $1.8 million."

Karen frowned. "Why replace it now and not in two years?"

This was the core question of determining the replacement timing. If you replace assets too early, you pay unnecessarily high interest and depreciation costs. Replacing too late leads to exponentially rising maintenance costs and unacceptable risks. The optimal moment is where the average annual costs are lowest. Karen realized that the choice to replace a few years earlier or later made a difference of tens of thousands of dollars per year.

11.1 Strategic Pressure from Management

At eleven o'clock, Karen sat across from director Hank in his office. "The power plant determines the continuity of our entire production," Hank began. "A wrong choice puts the plant's profitability at risk. We want systematic capital allocation with proof of return on every investment."

Karen had learned a crucial lesson about asset management: most costs are locked in when you spend the least money. The graph on her screen made this painfully clear — in the design phase you spend only 5% of the total life cycle costs, yet you determine 40% of all future expenditures.

"Look at this pattern," Karen said to the team, pointing to the curve. "By the time we're in the construction phase, we've already locked in 70% of the operational costs."
Hank looked at the numbers. "So designing cheap means expensive—" exploitation?"
"Exactly. If we save $50,000 now on a better design, it might cost us $500,000 extra over the next 20 years in energy, maintenance, and replacements."

11.2 Systematic Portfolio Approach

After her first conversation with Pete, Karen realized she was not looking at a single asset, but at a strategic decision that would affect the entire plant. The power plant was critical to all production processes, but formed part of a larger infrastructure system that
had to be fully optimized.

The traditional approach would be to look at the boilers in isolation and make a simple repair-or-replace decision. But Karen had learned from her earlier projects that this approach yields suboptimal results. Assets do not function in isolation, but as part of integrated systems where the performance of the whole is more important than that of individual components.
"Before we start looking at numbers," Karen said to her team, "we first need to get clear on exactly what we should be replacing. A one-to-one replacement of just the boilers with the same type is probably not the best option at all."
Pete nodded. "In 35 years, a lot of new technologies have come along. The current boilers were still designed for high capacity, while our demand varies much more."
"Exactly," Karen agreed. "And since we're going to replace it anyway, we might as well tackle a larger part of the power supply at the same time. What are the weakest links in our entire system?"
This systems approach opened up new possibilities that were not visible in a traditional asset-by-asset analysis. By expanding the scope from

boiler replacement alone to optimization of the complete energy system, strategic alternatives came into view that could create far more value.

The team developed four systematic alternatives, each representing a different philosophy. Running as-is meant accepting increased risk and rising maintenance costs in exchange for deferring capital investment. This option suited organizations that wanted to preserve cash flow and were willing to accept higher operational costs.

Renovation was an interim solution that addressed the most critical issues without complete replacement. For $180,000, critical components would be replaced and systems upgraded to modern standards.

Complete replacement meant renewal of the boilers for $1.8 million, which would implement modern technology and guarantee reliability for the next 25 years.

The integrated portfolio approach combined the boiler replacement with other planned investments to realize synergy benefits. This was the most complex option, but potentially also the most valuable.

Analysis of Alternatives

That afternoon, Karen had called her team together for a thorough analysis. They would analyze life cycle costs for all options over a fifteen-year period, explicitly taking into account the portfolio interactions between different assets — an extension of the LCC methodology from chapter 7.

The crucial difference from traditional LCC analysis was that she now had to quantify synergy effects between projects. Where individual asset analysis assumes isolated systems, portfolio analysis requires modeling interactions. Shared infrastructure, combined purchasing advantages, and operational synergies can only be identified when multiple assets are analyzed simultaneously.

The team developed a systematic approach in which each alternative was evaluated on four dimensions: direct project costs, operational impact over the service life, risk costs in the event of failure, and strategic value creation. This multidimensional evaluation went beyond traditional cost-benefit analysis by also incorporating non-financial aspects such as flexibility and future-proofing.

The analysis showed that complete replacement was economically superior to renovation or continued operation. But the real breakthrough came when the team began looking at integration opportunities with other planned investments.

11.3 Identifying Synergies Within the Portfolio

A week later, purchasing agent Sandra presented a fourth option. "We can integrate the new expansion of our CNC machining centers with the boiler replacement. The combined project delivers much more value than both projects separately."

This observation illustrated the core principle of portfolio management: synergies arise from the simultaneous execution of related projects. Where traditional project management treats each project as an isolated investment, portfolio management deliberately seeks out interactions that can create additional value.

The team identified three types of synergies that can be systematically leveraged. Infrastructure synergy arises when projects share similar facilities such as electrical, water, or process connections. Instead of duplicate investments, integrated systems can be designed that serve both projects.

Procurement synergy realizes cost advantages through combined purchase volume and supplier consolidation. A single supplier for related systems often offers better pricing, streamlined interfaces, and a consistent maintenance concept.

Operational synergy arises from more intelligent process control of integrated systems. Shared sensors, combined control, and optimized energy flows can yield efficiency advantages that are not achievable with isolated assets.

Controller Michael worked out these synergies in concrete value creation. "The combined approach delivers 60% more return than the sum of individual projects through systematic synergy exploitation."

11.4 Scenario Analysis for Robust Decisions

They implemented digital twin technology to analyze the complex integrated option. "For portfolio decisions, uncertainty modeling is crucial," he explained. "Individual projects have predictable risks, but portfolio interactions create new uncertainties that we need to model."

The digital twin approach required a fundamentally different methodology than traditional LCC analyses. Instead of point estimates, the team used distributions of possible values for critical parameters. Energy costs, demand growth, and technology performance were modeled as range values based on historical volatility and expert assessments.
By simulating 1,000 scenarios, the team was able to assess the robustness of various alternatives. This Monte Carlo approach revealed not only expected returns, but also the probability of negative outcomes and the spread of possible results. For strategic decisions, this risk profile is often more important than the average expected return.
"The integrated approach not only has the highest expected return," Pete concluded, "but also the most stable results under varying conditions. This makes it a robust choice for long-term value creation."

The analysis confirmed that portfolio management can deliver superior risk-adjusted returns through diversification within the project itself and through the flexibility that integrated systems offer under changing circumstances.

11.5 The Decision

For the presentation to management, Karen developed a structured business case following the methodology from chapter 8, but applied at portfolio level. The case systematically compared all four options: keep running ($-1.6 million), renovation ($+0.65 million), replacement ($+2.4 million), and integrated portfolio approach ($+3.8 million).
The business case demonstrated that the integrated option not only had the highest absolute return, but also the best risk-adjusted return. By leveraging synergies, the portfolio created more value than the sum of the individual projects.

Hank smiled after the presentation. "Approved. We're going with the integrated portfolio approach. This systematic approach gives me confidence that we're making the right decision."

The Results

Eighteen months after implementation, the results demonstrated the power of systematic portfolio management. The integrated project achieved 155% ROI and created $3.6 million in net present value. More importantly, the success transformed Metal Precision Group from reactive firefighting to strategic capital allocation.

"The culture has fundamentally changed," Karen reflected. "We now make investment decisions based on data and systematic analysis instead of intuition and crisis management."

This transformation extended far beyond the energy project. Ten other assets received the same portfolio treatment, identifying $2.8 million in additional value creation. Systematic capital allocation became standard for all major investments, transforming the organization toward proactive asset management.

11.6 Summary: From Methodologies to Portfolio Strategy

Karen's story illustrates how systematic portfolio management can transform asset decisions from reacting to problems into strategic value creation. By systematically analyzing four alternatives using digital twin modeling and LCC methodologies, her team created $3.6 million in value instead of the $2.4 million that traditional asset-by-asset analysis would have yielded.

The difference between portfolio management and the individual methodologies from earlier chapters lies in three dimensions.

First, the focus shifts from optimizing individual assets to optimizing the total system. Assets do not function in isolation; their value is determined by their role in the whole.

Second, synergies between projects are systematically identified and realized. Combined projects can create significantly more value than individual projects, but this requires portfolio thinking from the very beginning.

Third, scenario analysis with digital twin technology enables robust decision-making despite uncertainty about future developments.

The core of successful portfolio management lies in integrating all reliability engineering methodologies from this book into a strategic framework. FMEA identifies risks at the portfolio level. Design for maintainability is applied to integrated systems. RCM develops maintenance strategies for new configurations. LCC compares alternatives including synergy effects. Business cases justify strategic capital allocation. Supplier selection focuses on integrated solutions. Commissioning prepares organizations for new portfolio configurations.

For asset-intensive organizations, systematic portfolio management offers the opportunity to significantly improve their return on capital through strategic decision-making based on robust analyses. The difference between organizations that excel in asset management and organizations that struggle lies not only in the quality of their assets, but also in the quality of their decisions about their portfolio. By applying portfolio management systematically, organizations can transform from reactive troubleshooting to strategic value creation.

Epilogue: Engineering for Reliability in Perspective

Karen, Eighteen Months Later

Karen walked through the power plant she had commissioned eighteen months earlier. The dashboard showed numbers that made all the difference: 99.2% availability, 35% lower maintenance costs than comparable installations, not a single major failure.

"Nine times out of ten when my phone rings about this installation, it's a scheduled maintenance notification rather than a crisis," she said. "Even with earlier projects, we spent months working through teething problems."

But the eighteen months had also brought new challenges. The perfect maintainability she had built in only worked if the maintenance organization was well organized. The carefully developed RCM strategies were slowly degrading due to time pressure. The operators needed more training than expected.

"Designing systematically for reliability solves a lot," Karen said. "But I also saw where it stops working."

What Engineering for Reliability Does and Does Not Solve

Karen's transformation shows the power of systematic reliability engineering. You now know how to build reliability in from day one, which methodologies work and why the 80/20 rule means that design and construction are crucial.

But as Karen discovered, challenges remain:

Work organization: A perfectly maintainable installation is useless if work orders are not properly planned or urgent repairs always take priority over preventive maintenance.

Recurring failures: Some failures keep recurring despite a reliable design, because the root cause lies in operational processes or usage patterns.

Inventory management: FMEA reliably predicts which parts you need, but if those parts have a four-week lead time while you only have two weeks...

Assembly and execution: Even perfectly designed assets will fail if assembly is incorrect. Misalignment, improper torque values, or contamination during maintenance will undermine your design strategy.
Operator involvement: The most sophisticated built-in diagnostics will not work if operators do not know how to use them or are not motivated to take signals seriously.

Different Routes to the Same Destination

Karen started, like many engineers, with reliability engineering. She produced reliable assets and created support for further improvements. But it is not the only starting point.
Some organizations start with workflow control because their maintenance department is drowning in chaos and ad-hoc work. This works especially well when you have existing assets that need to perform better. It creates structure and overview, which makes later improvements easier.
Others start with structural failure reduction because one recurring problem is their biggest cost driver. Or with preventive maintenance, because they finally want to get a grip on their maintenance costs.

The Nine Disciplines for Structurally Reliable Assets

The challenges that remain — from work organization to preventive maintenance, from inventory management to troubleshooting — each call for their own systematic approach. This book series covers nine disciplines that together answer the question: "What must an organization do to make its technical assets perform as well as possible?"

1. Maintenance Management – The Overarching Framework
2. Planning and Scheduling in Maintenance– Efficient Organization of Work Activities
3. Troubleshooting Breakdowns– Effective Intervention in Acute Problems
4. Root Cause Analysis in Maintenance– Addressing Recurring Problems
5. Preventive Maintenance – Proactive Asset Care
6. Spare Parts Management – Optimal Inventory Control

7. Precision Maintenance– Flawless Installation and Overhaul
8. Engineering for Reliability – Reliability During Design
9. Autonomous Maintenance – Operator Involvement

The nine books follow a logical structure in which each book builds on knowledge from earlier parts. The implementation order in your organization may differ, however — start where the pain is greatest and where you can achieve results quickly. In the introduction of this book you will find a complete overview of all parts and how they relate to one another.

More information about the complete series
For a current overview of all books, additional information and ordering options, visit:

vadeo.nl/books.html

Every discipline you add makes the others more effective. They reinforce each other. Good work flow control ensures you have time for precision work. Preventive maintenance works better when operators detect deviations early. Reliability engineering prevents problems that you would otherwise have to address structurally. The order in which you implement them depends on what your organization needs most.

Your Next Step

You have now learned a methodology that works. Karen's results — and the thousands of organizations that apply similar approaches — prove that systematic reliability engineering transforms organizations.
The question is: do you stop here, or do you build on this?

For now: start with what you have learned in this book. Apply FMEA in your next project. Design for maintainability. Develop a solid business case. Measure your results. Celebrate your successes.

Karen's advice: "Start where the pain is. Measure results. And if you discover that other aspects are holding you back, there are systematic solutions for those as well. We started with reliability engineering on new projects and that worked for us. But I know companies that started with preventive maintenance on existing assets and were just as successful. It's not about the order, but about the systematic approach."

Appendix A Extended Scorecard for Supplier Evaluation

For the selection of suppliers of complex assets and systems

1. Technical Capacity (30% weighting)

Product Quality (10 points)	
10	Best quality in the market, zero defects track record
8	Excellent quality, minimal defects
6	Good quality, acceptable defect rate
4	Reasonable quality, some quality issues
2	Moderate quality, regular problems
0	Poor quality, frequent failures

Innovation Capacity (10 points)	
10	Market leader in innovation, >5% of revenue toward R&D
8	Strong innovation, 3-5% R&D investments
6	Reasonable innovation, regular new products
4	Limited innovation, primarily improvements
2	Little innovation, follows the market
0	No innovation, outdated technology

Integration Skills (10 points)	
10	Fully customizable, excellent system integration
8	Good adjustment capabilities
6	Standard adjustments possible
4	Limited flexibility
2	Little customization possible
0	Standard products only

2. Operational Reliability (25% weighting)

MTBF History (8 points)	
8	Proven MTBF >10 years for critical components
6	MTBF 5-10 years, good reliability
4	MTBF 2–5 years, acceptable reliability
2	MTBF 1-2 years, regular maintenance required
0	MTBF <1 year, frequent maintenance

Warranty claims percentage (8 points)	
8	<1% claims, transparent reporting
6	1-3% claims, good handling
4	3-5% claims, reasonable handling
2	5-10% claims, slow processing
0	>10% claims of opaque handling

Reference Performance (9 points)	
9	Excellent references with comparable clients
7	Good references, a few minor issues
5	Reasonable references, mixed experiences
3	Mediocre references, recurring problems
1	Poor references or refusal to provide them
0	No references available

3. Service & Support (20% weighting)

Local presence (7 points)	
7	Local service team, <4 hour response time
5	Regional service team, <24 hour response time
3	National service team, <48-hour response time
1	Limited local presence
0	No local service

Technical support (7 points)	
7	24/7 hotline with experts, remote diagnostics
5	Working days support with good expertise
3	Limited support, basic technical knowledge
1	Minimal support available
0	No technical support

Parts Service (6 points)	
6	Local stock, <24-hour delivery of critical parts
4	Regional inventory, 2-3 day delivery
2	Central inventory, 1 week delivery
0	No Guaranteed Availability

4. Financial (15% weighting)

Price Competitiveness (5 points)	
5	Best value for money in segment
4	Competitive pricing
3	Reasonable prices for the quality offered
2	Prices above market average
1	Expensive but acceptable for quality
0	Unacceptably expensive

Financial Stability (5 points)	
5	AAA credit rating, strong cash flow
4	Good credit rating, stable finances
3	Reasonable financial position
2	Some Financial Concerns
1	Financial problems but still operational
0	Serious Financial Problems

Total Cost of Ownership value (5 points)	
5	Lowest TCO Through High Reliability
4	Good TCO Value
3	Reasonable TCO
2	Higher TCO Due to Maintenance/Energy
1	High TCO but acceptable
0	Unacceptably High TCO

5. Strategic Fit (10% weighting)

Partnership Readiness (3 points)	
3	Real partnership mentality, capital investment willingness
2	Good teamwork, flexible
1	Standard supplier-customer relationship
0	Transactional attitude only

Technology roadmap (4 points)	
4	Clear innovation strategy, future-proof engineering
3	Good Development Plans
2	Basic roadmap present
1	Limited future plans
0	No clear technology strategy

Market Position (3 points)	
3	Market leader with strong competitive position
2	Strong market player
1	Reasonable market position
0	Weak market position or niche player

- **90–100: Preferred Partner** – Strategic choice despite higher price
- **80–89: Excellent supplier** – Highly recommended
- **70–79: Good supplier** – Acceptable with minor risks
- **60–69: Marginally acceptable** – Only with significant price advantages
- **50–59: Risky** – Not recommended for critical applications
- **< 50: Unsuitable** - Do not select

Must-Have Contract Provisions (for scores >70):

Performance Guarantees	98% availability first 2 years
Response Times	4 hours by phone, 24 hours on-site
Parts Warranty	15 years availability of critical parts
Software Support	Minimum 10 years of updates and patches
Penalty Clauses	Stepped System for Low Downtime
Escrow Agreement	It looks like you've provided a very short phrase without surrounding context. Here is the translation: For software and critical documentation

It looks like you've sent only a heading: '"Additional Requirements (for score >8.0)"' — this appears to already be in English, so no translation is needed.

If you have Dutch text you'd like me to translate, please paste it and I'll translate it right away.

Efficiency Guarantees	95% of design specification for 5 years
Proactive Monitoring	Remote Diagnostics and Predictive Maintenance
Training Program	For operators and maintenance personnel
Continuous improvement	Annual Evaluation and Improvement Plans

Evaluation process:

Pre-screening: Eliminate suppliers below minimum threshold (6.0)
Depth Evaluation: Complete Scorecard for Top 3-5 Candidates
Reference visits: For suppliers with a score >7.0
Financial due diligence: For final candidates
Contract Negotiation: Based on Weak Points in Score

Scoring team:

- Purchaser: Commercial and Contractual Aspects
- Engineer: Technical Assessment and Innovation
- Maintenance: Operational Certainty and Service

Finance: TCO Analysis and Financial Stability

This scorecard helps with objective supplier selection for complex assets where long-term performance and support are crucial for business continuity.

Figures

Figure 1-1 Expenditure and committed expenditure per life cycle phase 23
Figure 2-1 Schematic process flow of early equipment management..... 33
Figure 2-2 Impact of class on service life........ 37
Figure 2-4 Impact of number of components on service life........ 43
Figure 3-1 The consequences of a safety incident can be dramatic 54
Figure 3-2 Example of a risk matrix 56
Figure 3-3 Score of criticality ranking........ 58
Figure 5-1 Number of fitting sizes without (left) and with (right) standardization........ 86
Figure 6-1 Main decision diagram of RCM 97
Figure 6-2 Sub-decision diagram of RCM 98
Figure 7-1 Interest and depreciation in relation to the age of an asset . 116
Figure 7-2 Operational costs increase over the years........ 117
Figure 7-3 Optimal service life depends on many factors 118

Glossary — Terminology and Acronyms

5-Why Method A technique that asks 'why?' five times to move from symptom to root cause; effective for problems with a linear causal chain.

80/20 Rule of Reliability Engineering The principle that 80% of maintenance costs and reliability problems are determined during design and construction.

Accessibility The degree to which components are reachable for maintenance without disassembly of other equipment.

AI / Machine Learning Algorithms that recognize patterns in data to support decisions.

Asset A physical asset that creates value for the organization.

Availability The percentage of time that an asset is operationally available for production.

Benchmark A comparison of performance against industry standards or best practices.

Break-even Analysis Identification of the point where two alternatives have equal costs.

Built-In Diagnostics Sensors and systems that monitor the condition of components and provide early warnings.

Burn-in Testing Running systems under load for an extended period to eliminate early-life failures.

Business Case A structured justification of an investment including problem statement, solution, financial analysis, risk analysis, and recommendation.

CAPEX (Capital Expenditure) Investment costs that appear in accounting as interest and depreciation.

Class A measure of the perfection of the complete system, including quality of components, assembly, configuration, and operational context.

Cold Commissioning Testing all systems without process material.

Commissioning The process of converting a construction project into an operational plant with a focus on technology, business, and organization.

Condition-Based Maintenance (CBM) Maintenance based on the actual measured condition via inspection or monitoring, so that intervention only takes place when necessary.

Criticality Analysis A systematic method for prioritizing assets based on impact upon failure (safety, environment, production, quality, sustainability).

Detectability A score (1–10) for how easily a failure mode can be detected before it leads to problems.

Digital Twin A virtual replica of an asset or system for simulation and analysis without physical intervention.

Discount Rate The interest rate for converting future costs to present value.

Early Equipment Management (EEM) A project management methodology in which operators, technicians, and other end users are involved in design and implementation from the very beginning.

Edge Computing Processing power close to machines for real-time analysis without cloud latency.

Engineering for Reliability A systematic approach in which reliability is built into technical installations from the very first design.

Escrow Arrangement An arrangement whereby critical software and documentation remain accessible in the event of supplier bankruptcy.

Evident Failure A failure with immediately noticeable consequences.

Failure Mode The specific way in which a component or system can fail to perform its function.

FAT (Factory Acceptance Test) A test at the supplier's facility in a controlled environment for extensive reliability testing.

FMEA (Failure Mode and Effect Analysis) A systematic methodology for identifying potential failure modes, causes, and effects during the design phase.

Function Test A periodic test to verify whether hidden failures, especially in safety systems, have already occurred.

Functional Requirements Specifications that describe what the system must do, without defining how.

Gate-keeping A decision process at the transition between project phases with specific approval criteria.

Hidden Failure A failure that only has consequences when a second failure occurs (for example, a safety system failure).

Hot Commissioning Testing with actual process material under real operating conditions.

HSE (Health, Safety, Environment) Health, Safety, and Environment — the department responsible for compliance with requirements.

Industry 4.0 The fourth industrial revolution: digitalization, connectivity, and intelligent systems.

Infrastructure Synergy Value creation through sharing facilities such as electricity, water, or process connections.

Inverted Organizational Chart A concept whereby operators/end users have the most important role and management supports them.

IoT (Internet of Things) A network of sensors that continuously collect data on asset condition for real-time monitoring.

Kaizen A philosophy of small, continuous improvements.

KPI (Key Performance Indicator) A measurable indicator for assessing the performance of maintenance or assets.

LCC (Life Cycle Costing) A methodology for calculating total costs over the complete life cycle of an asset.

Life Cycle Cost (LCC) All costs of an asset from design to decommissioning: design costs, construction costs, operational costs, and decommissioning costs.

LMRA (Last Minute Risk Analysis) A systematic risk assessment performed immediately before the start of work activities.

LOTOTO (Lock Out, Tag Out, Try Out) A safety procedure for isolating energy sources before maintenance begins: locking out, tagging, and testing the isolation.

Maintainability The degree to which a system can be effectively and efficiently maintained or restored to its functional state.

Modularity Designing in interchangeable modules that can be replaced and maintained independently.

Monte Carlo Analysis Simulates thousands of scenarios for assessing the robustness of decisions under uncertainty.

MTBF (Mean Time Between Failures) The average time between consecutive failures of an asset.

Multidisciplinary Team A team with representatives from different disciplines (design, maintenance, production, quality, procurement).

Nine Elements of Downtime Administrative delay, logistical delay, access time, diagnosis time, obtaining parts, active repair time, test time, startup time, administrative closeout.

NPV (Net Present Value) The present value of future cash flows, discounted at the cost of capital.

OEE (Overall Equipment Effectiveness) A performance measurement consisting of Availability × Performance × Quality, expressed as a percentage.

Operational Readiness A systematic assessment of whether the organization is ready to take over a new installation (technical, personnel, procedures, maintenance, etc.).

Operational Synergy Efficiency gains through smarter process control and integrated systems.

OPEX (Operational Expenditure) Operational costs: personnel expenses, maintenance costs, energy costs.

Payback Period The time required to recover an investment through annual savings.

PDCA Cycle A continuous improvement method: Plan–Do–Check–Act.

PED (Pressure Equipment Directive) European directive for the safety of pressure equipment.

Portfolio Management A strategic approach in which multiple assets are optimized together for synergy benefits.

Predictive Maintenance A maintenance strategy based on predicting failures through data analysis.

Probability (Occurrence) A score (1–10) for the likelihood that a specific failure mode will occur.

Procurement Synergy Cost advantages through combined volume and supplier consolidation.

Project Phases with Approval Criteria Systematic verification that reliability requirements are met for progression to the next phase.

Quality The degree to which components meet their technical specifications.

RAMS Reliability, Availability, Maintainability, Safety — an integrated methodology for reliability from specification.

RCM (Reliability Centered Maintenance) A systematic methodology for developing maintenance strategies based on seven core questions about functions, failures, and consequences of failure.

Reliability The probability that an asset will meet all specified requirements without failure during a defined period of time under specific conditions.

Reliability Growth Systematically improving reliability during commissioning by finding and resolving early-life failures.

Remaining Useful Life Analysis Determining the optimal moment for replacing aging assets.

Risk Analysis Systematic evaluation of uncertainties and their impact on expected results.

Risk Matrix Acceptable frequencies of incidents for different risk types and severity levels, as established by management.

Risk Priority Number (RPN) Calculation: Probability × Severity × Detectability; used for prioritization of risk control.

ROI (Return on Investment) Return on investment calculated as a percentage, expressed as annual savings divided by the investment.

Root Cause Analysis (RCA) A systematic methodology for identifying and eliminating the fundamental causes of failures, rather than treating symptoms. The goal is to resolve problems permanently so they do not recur.

Run-to-Failure Maintenance (RTF) A deliberate strategy to allow an asset to fail and only then repair it, applied when prevention is not economically justified.

SAT (Site Acceptance Test) A test of the integrated system in the actual operating environment.

Scenario Analysis Evaluation of an investment under various possible future developments.

Scorecard Supplier Evaluation A systematic instrument for objective comparison of suppliers on all relevant criteria.

Sensitivity Analysis Testing how robust conclusions are against changes in assumptions.

Service Level Agreement (SLA) Contractual agreements on availability, response times, and service quality of an external supplier.

Severity A score (1–10) for the impact of the failure mode on safety, production, quality, or other aspects.

Stakeholder Analysis Systematic identification and involvement of all stakeholders in a project.

Standardization Use of uniform components, tools, and procedures for reduced complexity.

System Reliability The reliability of the complete system, where the number of components directly affects the probability of failures.

TCO (Total Cost of Ownership) All costs for owning and using an asset over its full life cycle.

Technical Capacity Evaluation of expertise, design, production capabilities, and innovation capacity of the supplier.

Technical Requirements Specifications that describe how the system must operate, including materials, standards, and configurations.

Time-Based Maintenance Periodic replacement based on a fixed time interval.

Track Record Proven performance of the supplier on comparable projects and clients.

Visibility The ability for technicians to assess the condition of components without disassembly.

Warranty for Performance A contractual commitment by the supplier regarding availability, reliability, and efficiency.

Acronyms and Abbreviations

AI Artificial Intelligence
CAPEX Capital Expenditure
EAM Enterprise Asset Management
EEM Early Equipment Management
FAT Factory Acceptance Test
FMEA Failure Mode and Effect Analysis
HSE Health, Safety, Environment
IoT Internet of Things
KPI Key Performance Indicator
LCC Life Cycle Costing
LMRA Last Minute Risk Analysis
LOTOTO Lock Out, Tag Out, Try Out
MTBF Mean Time Between Failures
MTTR Mean Time To Repair
NPV Net Present Value
OEE Overall Equipment Effectiveness
OPEX Operational Expenditure
PED Pressure Equipment Directive
RAMS Reliability, Availability, Maintainability, Safety
RCA Root Cause Analysis
RCM Reliability Centered Maintenance
ROI Return on Investment
RPN Risk Priority Number
SAT Site Acceptance Test
SLA Service Level Agreement
TCO Total Cost of Ownership
TPM Total Productive Maintenance

www.ingramcontent.com/pod-product-compliance
Lightning Source LLC
Chambersburg PA
CBHW050527240926
51846CB00003B/728
* 9 7 8 9 0 8 3 6 4 3 1 9 9 *